BIOTECHNOLOGY FOR WASTE AND WASTEWATER TREATMENT

BIOTECHNOLOGY FOR WASTE AND WASTEWATER TREATMENT

by

Nicholas P. Cheremisinoff, Ph.D.

np NOYES PUBLICATIONS
Westwood, New Jersey, U.S.A.

Library of Congress Catalog Card Number:
ISBN: 0-8155-1409-3
Printed in the United States

Published in the United States of America by
Noyes Publications
Fairview Avenue, Westwood, New Jersey 07675

10 9 8 7 6 5 4 3 2 1

Library of Congress Cataloging-in-Publication Data

Cheremisinoff, Nicholas P.
Biotechnology for waste and wastewater treatment / Nicholas P. Cheremisnoff.
p. cm.
Includes bibliographical references and index.
ISBN 0-8155-1409-3
1. Sewage--Purification--Biological treatment. 2. Water--Purification--Biological treatment. I. Title.
TD755.C547 1996 96-45255
628.3--dc21 CIP

PREFACE

This book examines the practices used or considered for biological treatment of water/wastewater and hazardous wastes.

The technologies described involve conventional treatment processes, their variations, as well as recent research. The book is intended for those seeking an overview of the field, and covers the major topics. The book is divided into five principal sections, and references are provided for those who wish to dig deeper.

Nicholas P. Cheremisinoff

NOTICE

ABOUT THE AUTHOR

Nicholas P. Cheremisinoff is a private consultant to industry, academia, and government. He has nearly twenty years of industry and applied research experience in elastomers, synthetic fuels, petrochemicals manufacturing, and environmental control. A chemical engineer by trade, he has authored over 100 engineering textbooks and has contributed extensively to the industrial press. He is currently working for the United States Agency for International Development in Eastern Ukraine, where he is managing the Industrial Waste Management Project. Dr. Cheremisinoff received his B.S., M.S., and Ph.D. degrees from Clarkson College of Technology.

CONTENTS

1 BIOTECHNOLOGY FOR INDUSTRIAL AND MUNICIPAL WASTES

Hazardous waste management remains the primary area of concern for many industries. Regulations, such as the Resource Conservation and Recovery Act (RCRA), the Toxic Substance Control Act (TSCA), and Superfund (CERCLA) as well as regulatory agencies, continue to keep corporate attention and the pressure on.

An important area of technology is biological treatment, popularly re-classified in recent years as **Biotechnology**. Biotechnology has its origins from an old science where we find applications in the antiquities. It is however a new technology under-going a resurgence in a wide range of applications, including past/present/future applications for the pollution engineer.

Natural decomposition of inorganic and organic materials has occurred for millions of years. Biological management of waste has been practiced for thousands of years. Most microorganisms in use are extracted from soil and water bodies and more recently technically developed for specific applications, and uses organic and toxic materials as sources of energy and carbon. While in the future, biological treatment will be based on microorganisms, a drastic departure from the past will most likely take place based on the new science of recombinant DNA.

The following are examples of recent research and applications of wastes and toxics, using biotechnology control:

- Some 20 different bacteria are said to be capable of breaking down polychlorinated biphenyls into water and carbon dioxide. One of these organisms from the genus Alcaliegenes is photoactivated by sunlight. Sunlight enhances the speed of degradation of PCB by some 400%.

- Researchers, involved in training bacteria--*Bacillus megaterium* and *Nocardiopsis*--to consume dioxin, observe that dioxin could easily penetrate the cell walls and be degraded faster if solvents such as ethyl acetate and dimethyl sulfoxide were added to the broth.
- A strain of genetically engineered microorganisms degrades 95% or more of the persistent 2,4,5-T within a week. Microbes can also degrade a variety of dichlorobiphenyls and chlorobenzoates.
- Scientists have isolated a strain of *Pseudomonas* that uses 2,4-D as a source of carbon. The gene involved was isolated and inserted in a different host bacteria.
- A number of microorganisms containing plasmids bearing genes for the degradation of aromatic molecules--toluene and xylene diverse salicylates and chloride derivatives of 4-chlorocatecol--have been tested.
- Formulation of bacterial mutants are commercially available for a variety of wastewater treatment problems. Specially formulated preparations are used for petroleum refinery/petroleum chemical plant wastewater cleanups. The bacteria degrades various hydrocarbons and organic chemicals (benzenes, phenols, cresols, napthalenes, amines, alcohols, synthetic detergents, petroleum (crude and processed)).
- Grease eating bacteria having successfully been used in cleaning clogged sewers.
- A major problem in recent decades has been the appearance of new chemicals in the environment stretching the ability of microorganisms to evolve by adaptation of existing catabolic enzymes or by the appearance of new metabolic pathways, the ability to degrade persistent xenobiotic compounds. We are constantly learning from such organisms and selecting those that show a maximum rate of biodegradation with maximum substrate utilization and minimum microbial biomass production.

Wastewater Treatment

Biological treatment is one of the most widely used removal methods as well as for partial or complete stabilization of biologically degradable substances in wastewaters and wastes. Suspended, colloidal or dissolved degradable organic material, quantities and ratios depend on the nature of the wastewater. Characteristics of wastewaters are measured in terms of Chemical Oxygen Demand (COD), Biochemical Oxygen Demand (BOD), and Volatile Suspended Solids (VSS).

Most biological waste and wastewater treatment processes employ bacteria as primary microorganisms; certain other microorganisms may play an important role. Degradation of organic matter is effected by its use as food by microorganisms to produce protoplasm for new cells during the growth process. Population dynamics of bacteria in biological treatment depends on environmental factors which include: pH; temperature; type and concentration of the substrate; hydrogen acceptor; essential nutrient concentration and availability; concentration of essential nutrients (e.g., nitrogen, phosphorous, sulfur, etc.); essential minerals; osmotic pressure; media toxicity; byproducts; and degree of mixing.

Metabolic reactions occurring within a biological treatment process can be divided into three phases:

- Oxidation
- Synthesis
- Endogenous respiration

Oxidation-reduction proceeds either in the presence of free oxygen (aerobically), or in its absence (anaerobically). Overall reactions may be different under aerobic/anaerobic conditions; microbial growth and energy utilization are similar. The three phases are:

Organic matter oxidation (respiration)

$$C_xH_yO_z + O_2 \rightarrow CO_2 + H_2O + \text{energy}$$

Cell material synthesis

$$C_6H_6O_3 + NH_3 + O_2 \rightarrow C_5H_7NO_2 + CO_2 + H_2O$$

Cell material oxidation

$$C_5H_7NO_2 \rightarrow NH_3 + 5CO_2 + 2H_2O + \text{energy}$$

Various conventional methods that are used in biological treatment are listed in Table 1 along with the treatment agents and typical wastes that are treated.

TABLE 1

METHODS OF BIOLOGICAL TREATMENT

Process	Treatment Agent(s)	Wastes Treated
Trickling filters	Packed bed (stones or synthetic) covered by microbial film	Acetaldehyde, benzene, chlorinated hydrocarbons, nylon, rocket fuel
Activated sludge	Aerobic microorganisms suspended in wastewater	Refinery, petrochemical and biodegradable organic wastewaters
Aerated lagoon	Surface impoundment plus mechanical aeration	Biodegradable organic chemicals
Waste stabilization ponds	Shallow surface impoundments plus aeration to promote growth of algae and bacterial and algal symbiosis	Biodegradable organic chemicals

BOD Removal

In wastewater treatment microorganisms are not present as isolated cells, but as a collection of microorganisms (such as bacteria, yeast, molds, protozoa, rotifers, worms and insect larvae) in a mass. These microorganisms tend to collect as a biological floc called biomass and generally possess good settling characteristics. Biological oxidation/stabilization of organic matter proceeds as follows:

- High rate of BOD removal from wastewater upon contact with active biomass. This removal and its extent depends on loading rate, waste type, and biomass.
- BOD is utilized in proportion to cell growth. Materials that concentrate on the biomass surface are decomposed by enzymes of living cells; new cells are synthesized; decomposition end products are washed into the water or escape into the atmosphere.
- Biological cell material oxidizes through endogenous respiration when food supply becomes limited.
- Biomass is converted to settleable material or removable solids.

Rates of these reactions depend on substrate transport rates, nutrients, and oxygen (in case of aerobic treatment). Any one or more of these transport rates can be the controlling factors that determine the process efficiency.

Types of Biological Processes

Biological treatment processes can be divided into three groups:

- **Aerobic stationary contact systems**--irrigation beds, irrigation sand filters, and trickling biomass remains stationary in contact with the solid support media (sand or rocks) and the wastewater flows around it.
- **Aerobic suspended contact systems**--the activated sludge process, its variations and aerobic lagoons comprise this group. In this group both biomass and substrate are in suspension or motion.

- **Anaerobic suspended contact systems**--anaerobic sludge digestion, anaerobic lagoons, and latter stages of landfills fall in this category.

Municipal Wastewater

Sewage is about 99.95% water and 0.05% waste. It is the spent water supply of a community. Due to infiltration of groundwater into loose sewer pipe joints, the quantity of wastewater is often greater than the water quantity that is initially consumed. More dilute sewage is a result of greater per capita water consumption, and industrial and commercial wastes contribute to sewage strength. Per capita sewage production can vary from less than 100 gallons per day for strictly residential areas to 300 gallons per day or more for industrialized areas. A typical sewage composition may be:

Total solids	600	mg/l	Mineral	20	mg/l
Suspended solids	200	mg/l	Filterable solids	400	mg/l
Settleable solids	120	mg/l	BOD (5 day 20°c)	54	g/cap./day
Colloidal solids	80	mg/l	Suspended	42	g/cap./day
Organics	60	mg/l	Dissolved	12	g/cap./day

The above estimate indicates a measure of the loading on a treatment plant (this may be additionally complicated by the presence of industrial effluents). The two principal processes utilized for biological (secondary) treatment are the trickling filter and activated sludge process.

Objectives in waste management change. Originally sewage treatment facilities were built primarily from a public health viewpoint but now include objectives such as oxygen protection for receiving waters. Clean water demand has increased more rapidly than population. This has given rise to the supply of complete treatment plants for small communities, developments, and isolated installations by manufacturers of waste treatment equipment in the form of packaged plants. A conventional scheme for wastewater treatment is illustrated in Figure 1. The pretreatment stage often consists of separating out coarse materials, grit, and oils. Primary treatment is comprised of the operations of flotation and sedimentation. Secondary treatment can be a combination of an activated sludge process, trickling filters, anaerobic or aerated

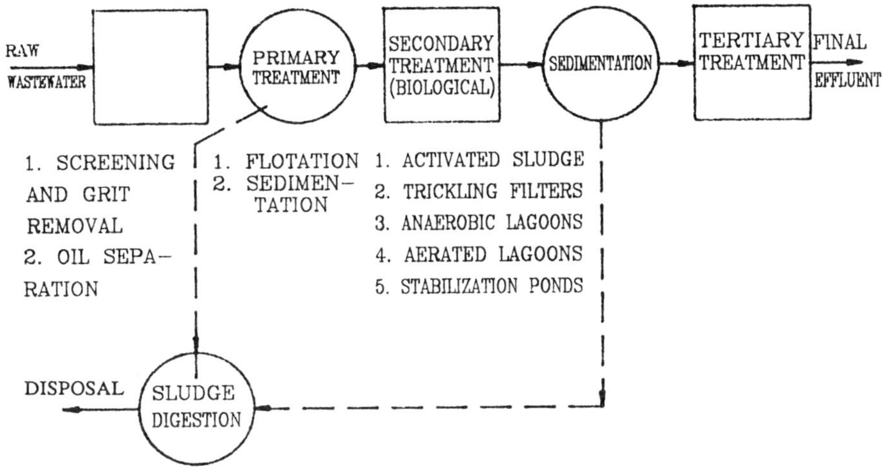

Figure 1. Typical wastewater treatment sequence.

lagoons, and stabilization ponds. This is often followed by sedimentation and then tertiary treatment, which is sometimes called "polishing."

Activated Sludge Process

The activated sludge process is a widely used and effective treatment for the removal of dissolved and colloidal biodegradable organics. It is a treatment technique well suited where organically contaminated wastewater exists. The activated sludge process is used by a wide range of municipalities and industries that treat wastewater containing organic chemicals, petroleum refining wastes, textile wastes, and municipal sewage.

The active sludge process converts dissolved and colloidal organic contaminants into a biological sludge which can be removed by settling. The treatment method is generally considered to be a form of secondary treatment and normally follows a primary settling basin. The flow diagram for a typical activated sludge treatment process is illustrated in Figure 2. There are several variations to this process including conventional arrangements, the contact stabilization process, and the step aeration process. Examples of these are given in Figure 3.

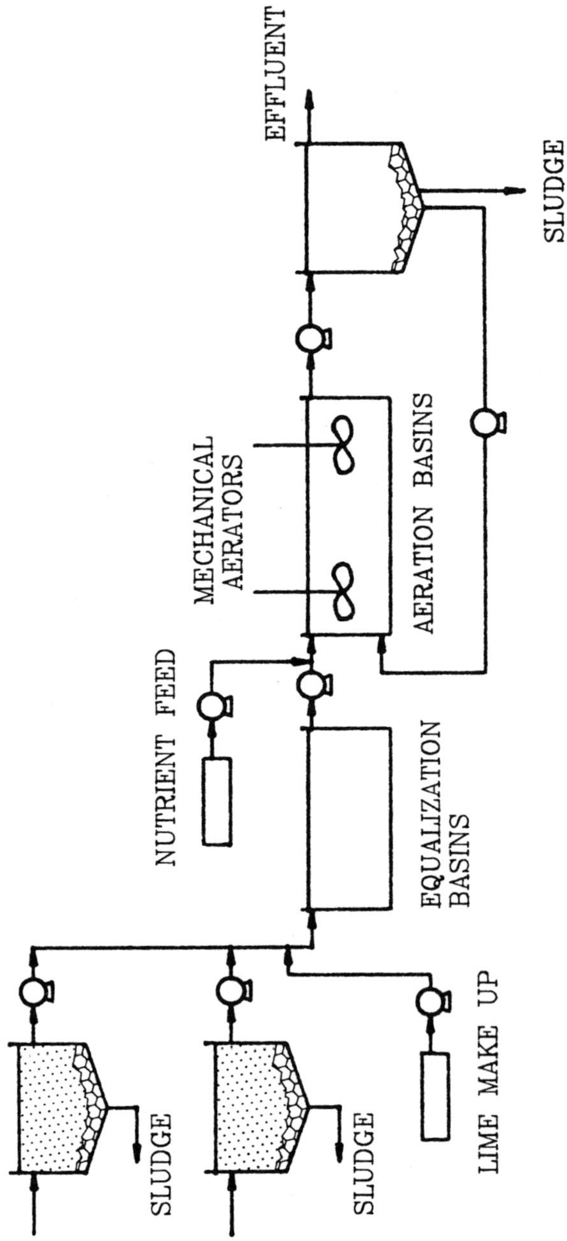

Figure 2. Activated sludge treatment flow diagram.

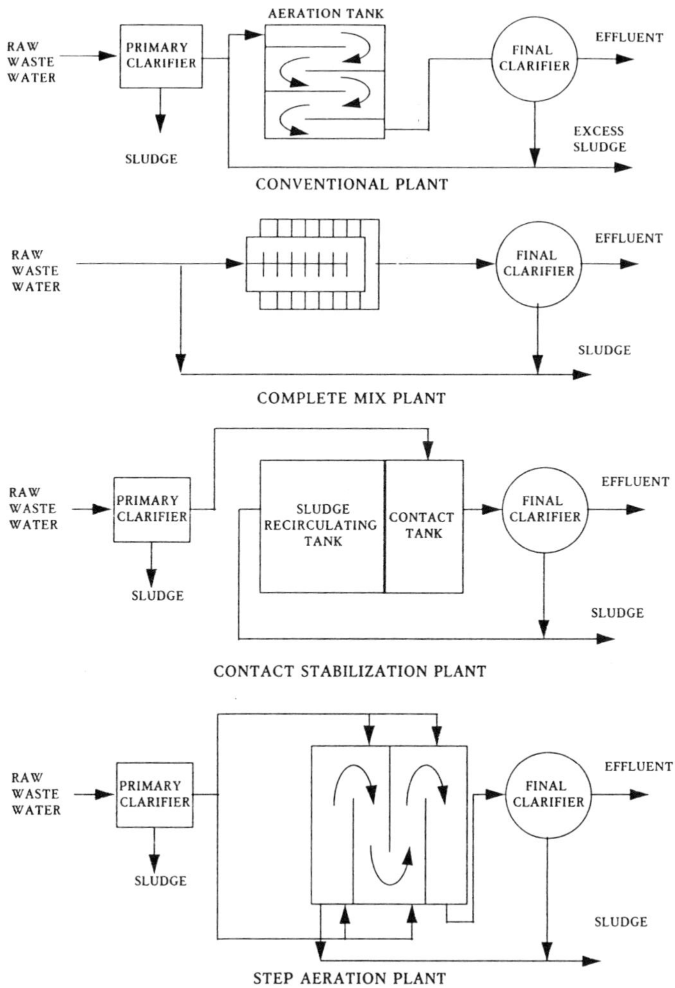

Figure 3. Variations of the activated sludge process.

In the activated sludge process the incoming wastewater is mixed and aerated with existing biological sludge (microorganisms). Organics in the wastewater come into contact with the microorganisms and are utilized as food and oxidized to CO_2 and H_2O. As the microorganisms use the organics as food they reproduce, grow, and die. As the microorganisms grow and are mixed together by the agitation of air, individual organisms floc together to form an active mass of microbes called activated sludge. The wastewater flows continuously into an aeration tank where air is injected to mix the activated sludge with the wastewater and to supply oxygen needed for microbes to breakdown the organic materials. This mixture of activated sludge and wastewater in the aeration tank is called mixed liquor. The mixed liquor flows from the aeration basin to maintain sufficient microbial population levels. This is the return activated sludge. The excess sludge which constitutes waste activated sludge is sent to sludge handling disposal.

Air is introduced into the system by aerators which are located at the bottom of the aeration basin, or by mechanical mixers (surface aerators). In addition, some processes utilize pure oxygen instead of air, known as pure oxygen activated sludge.

The microorganisms in activated sludge generally are composed of 70 to 90% organic and 10 to 30% inorganic matter. The microorganisms generally found in activated sludge consist of bacteria, fungi, protozoa, and rotifers. The growth and predominance of microorganism types are controlled by a number of circumstances including type of waste-organic matter (food), metabolic rate, and size. Predominance of certain microorganisms can be an indicator of treatment efficiency. Table 2 lists some of the microbes involved with the degradation of organic pollutants. There are variations to the conventional activated sludge process which are designed to overcome disadvantages inherent in specific applications. Some of these are described below.

Conventional (Plug Flow) Activated Sludge

The conventional activated sludge system is run in a plug flow pattern. That is, both the untreated wastewater and the return sludge are introduced at the head end of the aeration tank and mixed liquor is withdrawn at the opposite end. In an ideal plug flow system the flow will pass through the aeration tank without much mixing in the direction of flow. However, due to the aeration tank being aerated, mixing cannot be avoided. The best means of approaching plug flow conditions is to compartmentalize the chamber into a series of completely mixed reactors. A series of three or more reactors or compartments creates a truer plug flow design.

TABLE 2

MICROBIAL DEGRADATION OF VARIOUS ORGANIC POLLUTANTS

Pollutant	Microbes Involved
Petroleum hydrocarbons	200 + species of bacteria, yeasts, and fungi; e.g., Acinetobacter, Arthrobacter, Mycobacteria, Actinomycetes, and Pseudomonas among bacteria; Cladosporium and Scolecobasidium among yeasts
Pesticides/herbicides cyclodiene type (e.g., aldrin, dieldrin) organo-phosphorus type (e.g., parathion, malathion)	Zylerion xylestrix (fungus)
2,4-D	Pseudomonas, Arthrobacter
DDT	Penicillium (fungus)
Kepone	Pseudomonas
Piperonylic acid	Pseudomonas
Other chemicals:	
Bis (2-ethylhexyl)phthalate	Serratia marascens (bacteria)
Dimethylnitrosamine	Photosynthetic bacteria
Ethylbenzene	Nocardia tartaricans (bacteria)
Pentachlorophenol	Pseudomonas
Lignocellulosic wastes:	
Municipal sewage	Pseudomonas Thermonospora (a thermophilic bacterium)
Pulp mill lignins (various phenols)	Yeasts: Aspergillus Trichosporon Bacteria: Arthrobacter Chromobacter Pseudomonas Xanthomonas

Tapered Aeration

Plug flow processes are susceptible to shock loads. This is because the maximum concentration of flow is applied to microorganisms at the head end of the tank. Since a large oxygen demand is exerted at one location (the head end), adequate dissolved oxygen levels are difficult to maintain. The tapered aeration process is intended to deal with this problem, where a greater portion of the air is injected at the inlet end of the aeration tank where the greatest oxygen demand is required.

Step Feed Aeration

Step feed aeration is another variation of tapered aeration to equalize the oxygen supply and demand. Influent is fed at two or more points along the basin which equalizes the distribution of organic waste which subsequently results in more efficient oxygen use. The return sludge is returned to the head end of the tank where it initially does not come in contact with the raw wastewater. This reaeration assures that the sludge is not oxygen starved when it comes in contact with the waste and can readily absorb organic pollutants within a relatively short time. The aeration process also provides a short term reservoir for shock or toxic loads. The step aeration process can carry more solids under aeration than the conventional process: handles shock loads better; and has lower solids storage in the final settling tanks.

Contact Stabilization

Contact stabilization utilizes similar principles of sludge reaeration as discussed in the step-feed process. In this system the incoming wastewater is mixed briefly (20-30 minutes) with the activated sludge contact tank long enough for the microbes to absorb the organics but not actually long enough to break them down. The activated sludge is settled out and returned to another aeration/metabolization tank. The activated sludge is reaerated for 2-3 hours where the absorbed organics are oxidized. Following stabilization the reaerated sludge is mixed with incoming wastewater in the contact tank and the cycle starts again. Advantages of this process include: a smaller total aeration volume than the conventional processes, and as with the step feed process it can handle greater organic and shock loads due to the biological buffering capacity of the stabilization tank and lower solids inventory.

Complete Mix

In the complete mix system, the influent is fed as uniformly as possible along the entire length of the basin. As a result, the aeration tank is essentially homogenous resulting in uniform oxygen demand throughout the tank. This results in a homogeneous concentration of solids and substrates in the tank. This system is very stable and is less prone to toxic shocks which is a result of a relatively uniform population of organisms, and shock loads will be uniformly distributed to the tank and subsequently diluted.

Extended Aeration

The extended aeration process uses the same flow scheme as the conventional process but aerates the wastewater for 24 hours as opposed to 6-8 hours. Wastewater is aerated in a complex mix flow regime. This process operates in the endogenous respiration phase of the bacterial growth cycle in which there is not enough food remaining in the system to support all the microorganisms present because of low BOD_5 loading. The organisms are starved and undergo partial auto-oxidation utilizing their own cell structure for food. This results in a highly treated effluent and low sludge production. A disadvantage in this method is large oxygen requirements and tank volumes. Figure 4 illustrates the process of extended aeration activated sludge.

Oxidation Ditch

A variation of the extended aeration process is the oxidation ditch. In this system the wastewater is fed along a circular channel or racetrack and aerated by mechanical brushes or paddles along both sides of the channel. The typical oxidation ditch is 4-6 feet deep and is designed with a 24 hour retention time. A high degree of nitrification occurs due to the long retention time and high solids retention time (10 to 50 days). A flow diagram for the oxidation ditch process is illustrated in Figure 5.

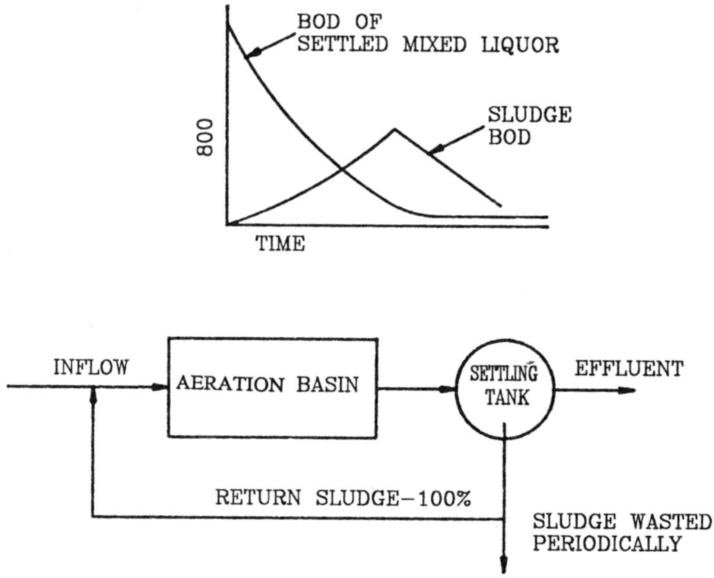

Figure 4. Extended aeration activated sludge.

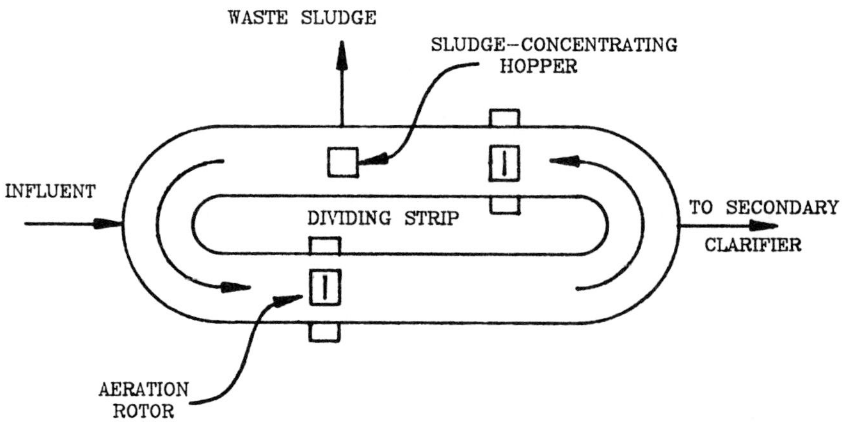

Figure 5. Oxidation ditch flow diagram.

Anaerobic Digestion

Major applications of anaerobic digestion are in the stabilization of concentrated sludges produced from the treatment of wastewater and in the treatment of some industrial wastes. The digestion is a complex biochemical process in which several groups of anaerobic and facultative organisms simultaneously absorb and break down organic matter. It can be described as a two-phase process:

- Facultative, acid-forming organisms convert the complex organic substrate to volatile organic acids. Acetic, propionic, butyric, and other organic acids are formed. Little change occurs in the total amount of organic material in the system, although some lowering of pH results.
- Second phase involves conversion of the volatile organic acids to principally methane and carbon dioxide.

The anaerobic process is essentially controlled by the methane-producing bacteria. Bacteria grow at a relatively low rate and have generation times which range from slightly less than 2 days to about 22 days. Methane formers are very sensitive to pH, substrate composition, and temperature. If the pH drops below 6, methane formation stops, and there is no decrease in organic content of the sludge. The methane bacteria are highly active in the mesophilic and thermophilic ranges. The mesophilic range is 79-110°F (26-43°C) and the thermophilic range is 113-149°F (45-65°C). Essentially all digesters in the United States operate within the mesophilic range. Table 3 illustrates the biochemical reactions occurring in the anaerobic digestion process. Anaerobic sludge digestion is a continuous process. Fresh sewage sludge is added continuously or at frequent intervals. The water separated from the sludge (supernatant) is normally removed as the sludge is added. Digested sludge is removed at less frequent intervals. Gas formed during digestion is removed continuously.

Stabilization of sludge by anaerobic digestion results in the production of methane gas which is insoluble in water and escapes as a gas. Thus, if no methane gas is produced there can be no waste stabilization. It is important to note that no waste stabilization occurs in

TABLE 3

THE ANAEROBIC DIGESTION PROCESS

Item	Composition
Raw Sludge	Complex substrate of carbohydrates, fats, and proteins
Microorganisms "A"	Principally acid formers
Nonreactive Products	CO_2, H_2O, stable and intemediate degradation products, cells
Reactive Products	Organic acids, cellular and other intermediate degradation products
Microorganisms "B"	Methane fermenters
Other End Products	H_2O, H_2S, cells and stable degradation products

the first stage, but occurs entirely in the second stage of the anaerobic digestion process. The anaerobic digestion process can be carried out in an airtight reactor. The standard-rate and high-rate systems are the two main digestion processes employed. In practice, four types of systems have evolved from the two basic digestion methods. They are:

- Standard-Rate Digestion, Single-Stage.
- High-Rate Digestion, Single-Stage.
- Two-Stage Process.
- Anaerobic Contact Process.

In the standard-rate, single-stage digestion process (refer to Figure 6), the contents of the digester are usually unheated and unmixed. Detention times vary from 30 to 60 days. In a high-rate digestion process, the contents of the digester are heated and completely mixed. The required detention time is 15 days or less. The primary function of the second stage is to separate the digested solids from the supernatural liquor. However, additional digestion and gas production may occur. Sludge digesters currently in use in the United States fall into one of four designs:

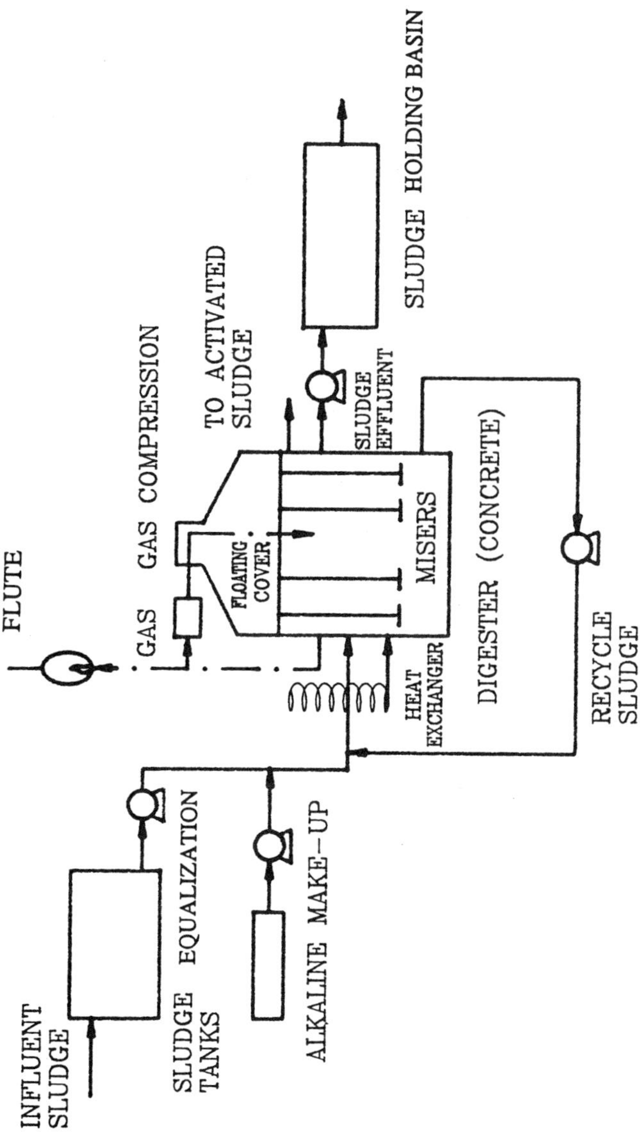

Figure 6. High-rate anaerobic digestion flow diagram.

1. Conventional Rate Digestion - One Stage
   a. heated or unheated
   b. detention time 30-60 days
   c. solids loading 0.03 - 0.1 lb VVS/ft^3 day
   d. intermittent feeding and withdrawal

2. High-Rate Digestion - One Stage
   a. heated 85-95°F (mesophilic range)
   b. detention time 15-20 days
   c. solids loading 0.1-0.2 lb VSS/ft^3 day
   d. continuous feeding and withdrawal
   e. process feature: homogeneity
   f. process feature: stratification

3. Two-Stage Digestion: a combination of Designs 1 and 2 above.

4. Anaerobic Contact Process: similar to Design 3 except sludge from the second stage is recycled to the head of the first stage.

Examples of these processes and digesters are illustrated in Figures 7 and 8.

MUNICIPAL TREATMENT PLANT SLUDGES

Wastewater sludge is being generated in enormous quantities at sewage treatment plants, particularly at activated sludge facilities. Regulations have mandated both the end of ocean dumping of sludges and provisions for full secondary treatment. These regulations result in increased production of sludge and the necessity to treat and dispose of it in an acceptable manner. Anaerobic digestion is one of the processes employed in the stabilization of these sludges, to remove from the raw sludge its odor, pathogens, putreseibility, and other offensive characteristics. The quantities of sludge produced in municipal operations as an example are considerable. Some typical volumes of sludges generated are reported in Table 4. There are various unit operations that are used in a typical sludge treatment process. The conventional design is shown schematically in Figure 9. The origins of sludges derived from wastewater treatment operations can be readily identified from Figure 10.

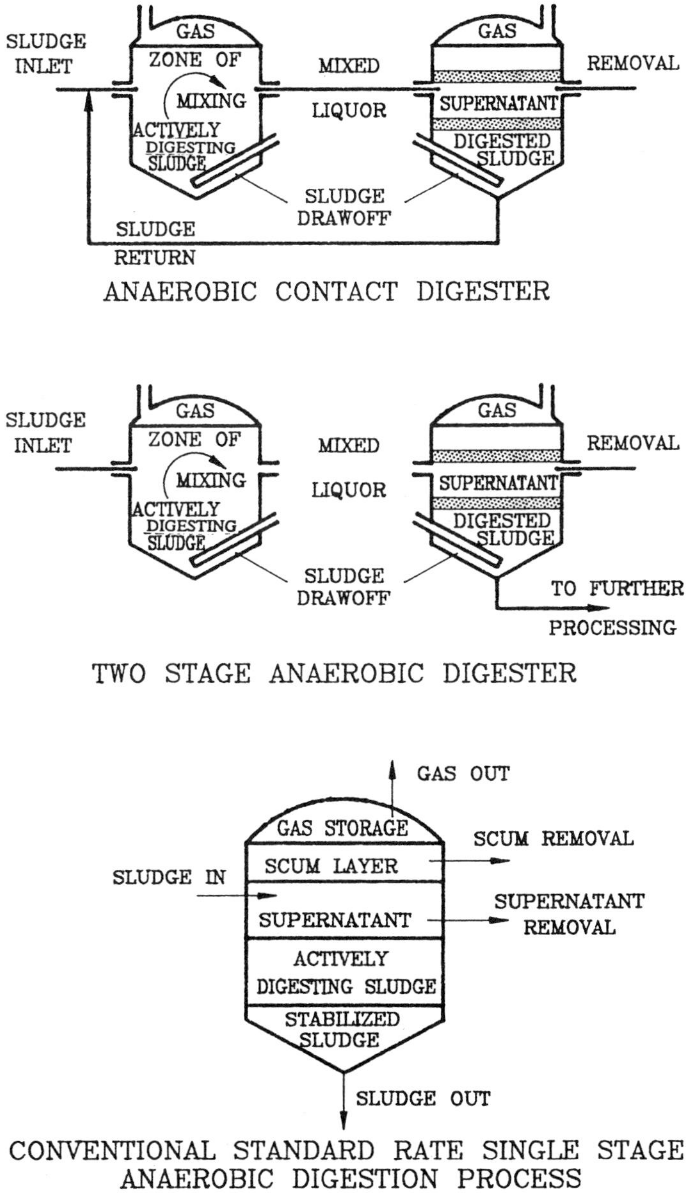

Figure 7. Sludge digesters used in the United States.

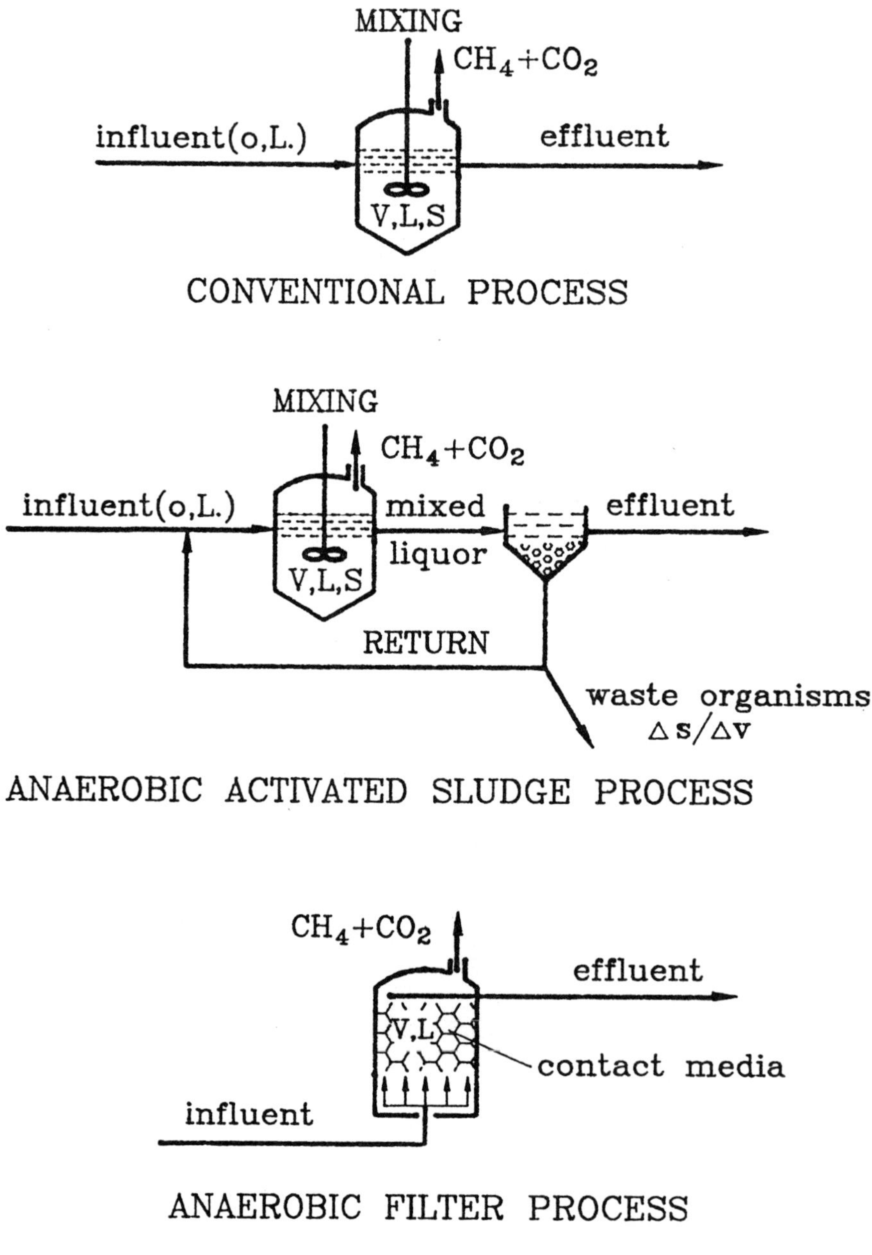

Figure 8. Anaerobic process designs.

TABLE 4

TYPICAL SLUDGE VOLUME PRODUCED BY CONVENTIONAL TREATMENT PROCESS

Process	Gallons Sludge/Million Gallons Waste
Primary Sedimentation	2,400 - 3,500
Trickling Filter	530 - 750
Activated Sludge	14,600 - 19,400

Using the methane produced by the anaerobic digestion of sludge to supply power for the sewage treatment plant is not new. This practice was used in the 1940s and 1950s by many municipalities, but was generally abandoned in the 1960s when electricity became inexpensive. Aerobic digestors were chosen in place of anaerobic ones because of their relative ease of operation and resistance to upset. However, they are energy users. Energy costs have resulted in a renewed interest in anaerobic sludge digestion since the methane produced can often heat the digester and supply the bulk of the power needed by the entire treatment plant.

Desulfurization

The bacteria, expert at mineral leaching can be applied to water decontamination in conjunction with other microorganisms. In many mining operations, water is pumped out of the mines to prevent flooding. Water used in milling processes becomes laden with soluble inorganic ions. Mine drainage from abandoned mines is loaded with a variety of metal salts. The practice has been to evaporate this water in holding ponds or to neutralize the acid flow and precipitate the metals with lime.

Many organisms have the ability to concentrate, accumulate, or precipitate metals allowing the recovery of elements of economic importance. Many bacteria are known to concentrate potassium, magnesium, manganese, iron, calcium, nickel, and cobalt. Other bacteria produce complexing agents which selectively extract metals from dilute solutions. Algae concentrate silica and green-brown algae and fungi concentrate zinc and other heavy metals. Mosses and higher plants concentrate mercury, nickel, zinc, uranium, cesium, and strontium.

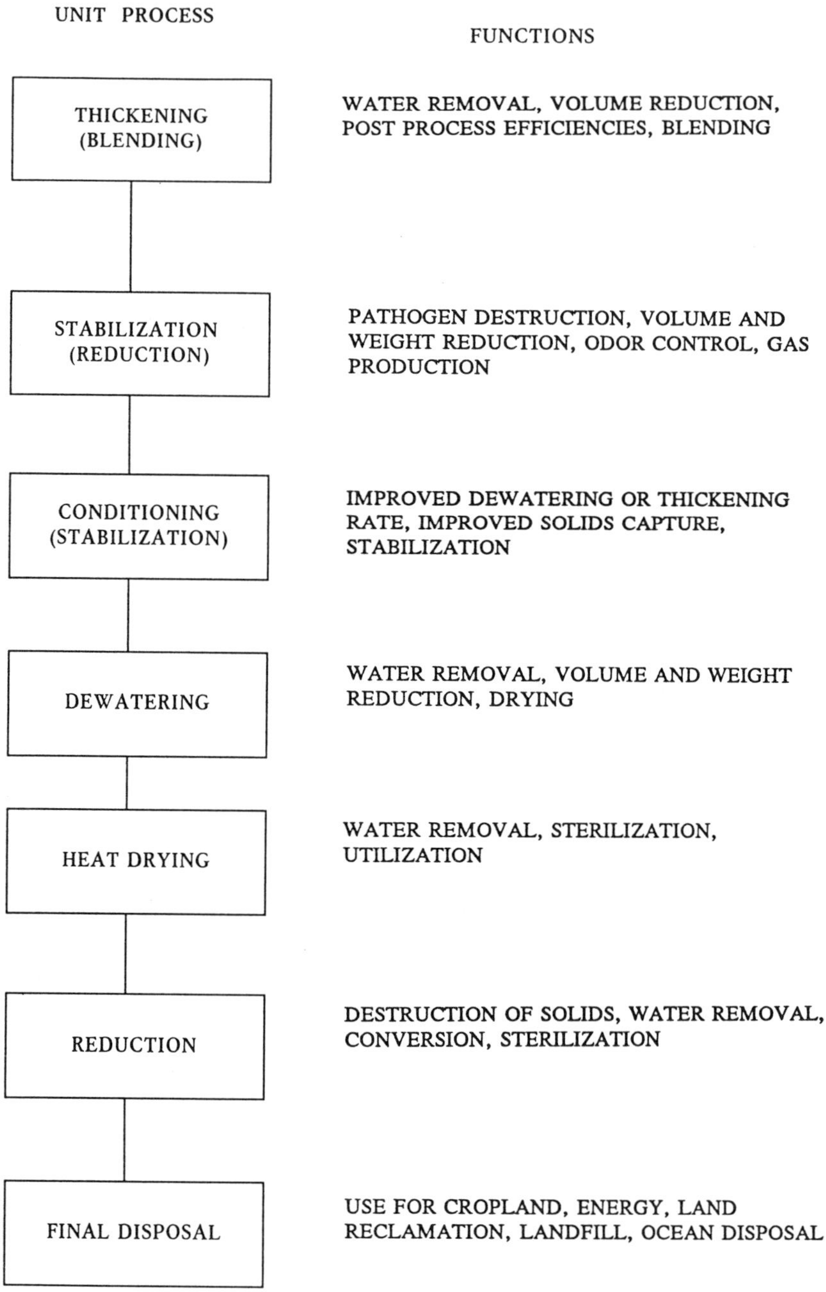

Figure 9. Sludge treatment processes and their functions.

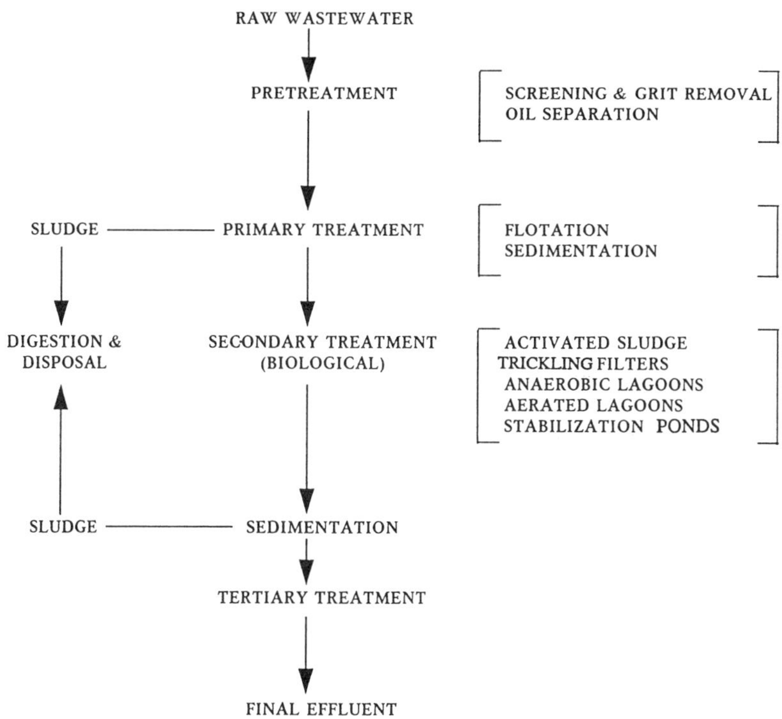

Figure 10. Typical wastewater treatment sequence incorporating biological treatment processes.

Sulfate reducing bacteria of the *Desulfovibrio*, *Desulfotomachulum*, *Desulfobacter*, *Desulfococcus*, *Desulfonema*, and *Desulfosarcina* general are especially adept at metal removal from water by producing hydrogen sulfide which precipitate these metals. The constituent members of these groups embrace a wide range of salinity or osmotic pressure, temperature, hydrostatic pressure, pH, Eh, and other environmental conditions.

These organisms have been put to work in mine wastewater cleanup operations. Settling ponds inoculated with sulfate reducing bacteria (*Desulfovibrio*, *Desulfotomaculum*), and sulfur oxidizing thiobacilli (*Thiobacil thiooxidants*, *T. thiopiaus*, *T. dentrificans*), and algae (*Chara*, *Spirogyra*, *Oscillatoria*) have been effective in lowering the concentration of uranium, selenium, and molybdenum in wastewater.

A variety of metallurgical effluents contain high concentrations of sulfate ions. A number of microorganisms can utilize this sulfate and convert it to an insoluble, stable non-leachable form. *Desulfovibrio* reduces sulfate to sulfide. *Chlorobium* and *Chromatium* photosynthetically oxidize H_2S to elemental sulfur. A mutualism between these bacteria is proposed.

$$SO_4^{-2} = \xrightarrow{\text{Desulfovibrio}} H_2S \qquad H_2S \xrightarrow[\text{Chromatium}]{\text{Chlorobium or}} S^{\circ}$$

A series of tests applied to solvent extraction raffinate demonstrates that a gas purged mutuallistic system of *Desulfovibrio* and *Chlorobium* can be used for the efficient conversion of sulfate to elemental sulfur. The extraction of minerals from ores, its benefitiation to a high quality material, and its fabrication into a useful product are all sources of very toxic materials. Many industrial wastes contain valuable metals diluted in a large mass or volume. Processes must be developed to simultaneously extract these valuable metals and reduce the attack on the environment.

Tests have been conducted using *T. ferroxidans* and *T. thiooxidants* to extract economically interesting metals from wastes:

1. Jarosite--a residue which accumulates during zinc production.
2. Sulidic dust concentrates from copper processing.
3. Fly ash from apyrite-roasting process.
4. Slag from a lead smelting process.

The results indicate that it is possible to stimulate bacteria already to work on these waste dumps to leach into solution economically valuable metals.

Flotation is probably the most common unit operation in metallurgical operations. One of the problems is that the reject water is laden with flotation chemical agents. Laboratory experiments have tested the ability of *Escherichia coli*, *Proteus retigerii*, *Klebsiella pneumoniae* and *Pseudomonas aerucinosa* to biodegrade sodium hexadecyl-sulfate, sodium oleate, and dodecylamine acetate. *Klebsiella* and *Proteus* appear to be the most efficient organisms. They handled very well the sodium hexadecyl sulfate, manage with the sodium oleate but had difficulty with the amine. Pseudomonas fluorescens has been used to remove flotation agents from wastewaters.

Nitrification/Denitrification

The nitrogen cycle is complex like all biogeological cycles. Its essential components are:

1. Nitrogen fixation

$$N_2 + H^+ \xrightarrow{\text{Rhizobium}} NH_3$$

2. Nitrification

$$NH_3 + O^2 \xrightarrow{\text{Nitrosomas}} NO_2^-$$

$$NO_2^- + O_2 \xrightarrow{\text{Nitrobacter}} NO_3^-$$

3. Denitrification

$$NH_3 + NO_3^- = \xrightarrow{\text{Pseudomonas}} H_2O \text{ or } N_2$$

During the biological denitrification, bacteria reduces nitrite or nitrate to nitrogen gas. The heterotropic bacteria require a source of carbon. In denitrification reactions, this supplemental source is often provided by methanol. The nitrogen reaction is a respiratory process encompassing the following reactions.

$$NO_3 + 1/3\ CH_3OH + 2/3\ CO_2 \longrightarrow NO_2 + CO_2 + 2/3\ H_2O$$

$$NO_2 + 1/2\ CH_3CH + 1/2\ O_2 \longrightarrow 1/2\ N_2 + 1/2\ CO_2 + 1/2\ H_2O + OH + 1/2\ CO$$

$$NO_3 + 5/6\ CH_3OH \longrightarrow 1/2\ N_2 + 5/6\ CO_2 + 7/8\ H_2O + H_2O$$

Biological reactions have been used in the conversion of ammonia-nitrogen to nitrate-nitrogen with attendant reduction in chemical oxygen demand and total organic carbon in coal gasifier effluents and municipal wastewaters.

Nitrogen, in its various forms, can deplete dissolved oxygen levels in receiving waters, stimulate aquatic growth, exhibit toxicity toward aquatic life, affect chlorine disinfection efficiency, present public health hazards, and affect the suitability of wastewater reuse. Nitrogenous materials enter the aquatic environment from natural or man-caused sources. Natural sources include precipitation, dustfall, non-urban run-off, and biological fixation. Activities that may increase quantities of nitrogen added to the aquatic environment are from fertilization of agricultural land and combustion of fossil fuels. Other man-related sources include urban and livestock feedlot run-off, municipal wastewater effluents, and subsurface drainage wastes. The average concentrations of nitrogen from natural sources is difficult to estimate but range from 0.02 mg/l to 0.2 mg/l. Nitrogen concentrations in raw municipal wastewaters are well documented, and values range from 15 to 50 mg/l of which approximately 60% is ammonia nitrogen, 40% is organic nitrogen, and a negligible amount (1%) is nitrite and nitrate nitrogen. Nitrogen concentrations of other man-related sources vary widely depending on the source. Treatment of these and other non-point sources is difficult if not impossible to treat. For the purpose of this control, methods can be divided into three broad categories:

- Biological methods of removal.
- Chemical/physical methods of removal.
- Other methods of removal.

Biological methods include nitrification in suspended growth; and attached growth systems: using trickling filters, rotating biological contractors, and packed bed reactors. Biological denitrification in both suspended and attached growth reactors is a developing method. Chemical physical methods for removal include breakpoint chlorination and ozone treatment, selective ion exchange, and ammonia stripping. In addition, other methods such as aquatics have been discussed.

The term nitrification is applied to the reaction in nature of the biological oxidation of ammonium (NH_4) first to the nitrite (NO_2), then to the nitrate (NO_3) form. The conversion of ammonium to nitrate is caused by the bacteria Nitrosomonas as follows:

$$2NH_4 + 3O_2 \xrightarrow{\text{bacteria}} 2NO_2 + 4H + 2H_2O$$

The nitrate in turn are oxidized by the bacteria *Nitrobacter* according to the following:

$$2NO_2 + O_2 \xrightarrow{\text{bacteria}} 2NO_3$$

The overall nitrification reaction is as follows:

$$NH_4 + 2O_2 \xrightarrow{\text{bacteria}} NO_3 + 2H + H_2O$$

For this reaction to go to completion about 4.6 mg/l of oxygen are required per gram of ammonia nitrogen. This oxygen demand is often identified as the nitrogenous oxygen demand (NOD).

The problem with nitrogen in wastewater is related to the NOD if ammonia nitrogen is discharged into the environment. This problem can be mitigated by either converting the ammonia and organic nitrogen to nitrate or by eliminating the nitrogen in the wastewater. Biological methods of nitrogen removal are classified under two broad categories, biological nitrification and denitrification.

Nitrification

The application of biological nitrification in municipal wastewater treatment is applicable to cases where ammonia removal requirements exist, without the need for complete nitrogen removal. Nitrification systems are broken up into combined carbon oxidation-nitrification and separate stage nitrification processes and each can be divided further into suspended and attached growth processes. Suspended growth processes are those which suspend the biological solids in a mixed liquor by a mixing mechanism. A subsequent clarification stage is required for returning these solids to the nitrification stage. Attached growth processes, on the other hand, retain the bulk of the biomass on the media and, therefore, do not require a solids separation stage for returning the solids to the nitrification reactor.

Attached Growth Processes--The trickling filter, rotating biological contactor, and the packed bed reactor are principal biological methods of nitrification. The trickling filter and rotating biological contactor can be considered for combined carbon oxidation-nitrification and separate stage nitrification, whereas the packed bed reactor is only considered as a separate stage process.

Trickling Filter (combined oxidation/nitrification)--The trickling filter consists of a bed of highly permeable media to which microorganisms are attached and wastewater is percolated. The filter media usually consists of rock, packing, etc. one to four inches in diameter to a depth of three to eight feet. Some filters have been built with plastic or redwood media and to depths of 15 to 25 feet.

The development and maintenance of nitrifying organisms in a trickling filter depends on factors such as organic loading and temperature. Kinetic theory for combined carbon oxidation nitrification has not been developed, therefore, the approach applied to date is largely empirical and relys mostly on specification of an organic loading rate suitable for application to each media. A typical loading rate for rock media to attain 75% nitrification is 10 to 12 lb BOD_5/1000 gal/day. At higher loading rates, such as that above 30 to 40 lb BOD_5/1000 gal/day, very little nitrification occurs. The specific media surface selected has an effect on the allowable organic loading. Greater specific surface in the media allows greater biological film development and a greater concentration of organisms within a unit volume.

Trickling Filter (separate stage nitrification)--This system initiated the development of the two-stage suspended growth system for nitrification. Later it was observed that nitrification occurred as a side affect of the original use of the two-stage trickling filters. Initially, the two-stage trickling filtration process was developed to increase the removal of organics in the effluent of high rate trickling filters. In separate stage nitrification applications, the rate of nitrification is proportional to the surface area exposed to the liquid being nitrified. Plastic media is commercially available with specific surface available from 27 ft^2/ft^3 to 68 ft^2/ft^3 and as such is the best alternative for a two-stage suspended growth system because of its high specific surface area. Pilot plant studies indicate that with temperatures of 13 to 19°C and influent NH_4-N of between 8 to 18 mg/l NH_4-N can be reduced to

2.5 mg/l for 4000 ft²/lb NH_4-N oxidized/day. Whereas with temperatures of 7 to 11°C and the same influent ranges, removal rates only averaged 3.0 mg/l per 4000 ft²/lb NH_4-N oxidized/day. When effluent levels of less than 2.5 mg/l NH-N are desired, consideration should be given to breakpoint chlorination for reducing the ammonia residual rather than increasing the surface area of the filter since a surface area of 10,000 to 12,000 ft²/lb NH_4-N oxidized/day will be required to reduce ammonia effluents to between 1 to 2.5 mg/l. Tests using rock media have shown the NH_4-N oxidation is only 15 to 50% of the plastic media rate. The main reason for this is rock media's lower specific surface area. A 1:1 recirculation ratio is normally considered adequate for this type of system. Effluent clarification is not required in separate stage trickling filter systems since organisms are attached to the media and solid levels are low. In pilot plant studies, it was found that the same low influent solids of 9 to 28 mg/l were evident in the effluent.

Rotating Biological Contactors (combined carbon oxidation/nitrification)--The rotating biological contactor or disk (RBD) has seen limited use in the United States in combined carbon oxidation/nitrification. The RBD process consists of a series of closely spaced circular disks of polystyrene or polyvinyl chloride. The disks are slowly rotated and submerged in the wastewater. In operation, biological organisms become attached to the surface of the disks and in about 10 days to two weeks, form a slime layer of aerobic biomass. In rotation, the disks biomass contact wastewater and then the atmosphere for absorption of oxygen. Disk rotation is the mechanism used for the removal of the excess solids from the disks by the shearing forces as the disks pass through the liquid. The treated effluent is mixed and in suspension, and the wastewater flow carries the suspension out of the disk section into a secondary clarifier for separation and disposal. Disk units are normally housed to avoid temperature drops, prevent algae growth on disk surfaces, and to protect from weather.

As a RBD system operates in series, organic matter is removed in the first disk stages and the subsequent disk stages are used for nitrification. Nitrification only occurs after the bulk of the BOD_5 has been oxidized. After the low levels of BOD_5 have been reached, the disk is no longer dominated by heterotrophs, and nitrification can proceed.

Rotating Biological Contactors (separate stage nitrification)--In order to obtain separate stage nitrification, the wastewater influent must be pretreated to remove organic carbon. As such, separate stage nitrification is very adaptable to upgrading existing secondary sewage treatment systems to meet low ammonia nitrification effluents. The BOD_5/TKN ratio must be sufficiently low to ensure a significant fraction of nitrogen to biomass. Some alternative pretreatment systems used are chemical treatment in activated sludge, roughing filters, and trickling filters. In addition to the separate stage trickling filter, Rotating Biological Contactors are very adaptable to separate stage nitrification.

The purpose for RBD combined carbon oxidation/nitrification may be applied to separate stage nitrification. The only difference is if the secondary effluent being treated has a BOD_5 or suspended solids content of less than 20 mg/l, the secondary clarifier may be eliminated. The most common RBD systems used is a four-stage system having four rows of disks. But the more stages, the greater the nitrification.

Packed Bed Reactors (PBR) (separate stage)--A packed bed reactor consists of a bed of media where nitrifying biological mass is developed, overlaying a support media and inlet chamber to which wastewater influent is introduced through an upflow configuration. Media bed types that have been used are anthracite coal, silica sand, stones, gravel, and a plastic packing media. Both air and high purity oxygen have been employed as a means of supplying the necessary oxygen for nitrification. Air can either be injected directly into the inlet chamber or distributed across the PBR floor. Oxygen can be bubbled directly into the PBR. Only monthly backwashing of light density anthracite or plastic media is considered necessary due to turbulence developed in the bed. Experience has shown that back-washing of a gravel media reactor is required on a more frequent basis, perhaps as much as once per day. Oxidation rates for a typical reactor range from 4 to 27 lb NH_4-N oxidized per 1000 ft^3/day. A typical reactor using crushed coal and loosely packed plastic can reduce average ammonia nitrogen concentration from 15 mg/l to less than 1 mg/l. This is at an ammonia loading of 6.9 lb NH_3-N/day/1000 ft^3 of media volume, detention time of 3.25 hours, operation temperature of 10°C. Temperature has a strong affect on the PBR process by increasing the temperature of liquid from 10°C to 20°C; the detention can be cut to less than two hours to obtain the same effluent characteristics.

Suspended Growth Process--The activated sludge process is the main suspended growth biological method nitrification used to treat sewage. Basically, there are two methods considered for nitrification by activated sludge systems: combined carbon/nitrification, accomplished by nitrification in the same tank where carbonaceous removal is accomplished; and separate stage nitrification which has developed to isolate the carbonaceous removal and the nitrification processes so that each could be separately controlled and optimized.

Not all of the modifications to the activated (combined carbon oxidation/nitrification) process are appropriate for nitrification applications. Some of the more successful applications are the complete mix plant, conventional plant, extended aeration plant, contact stabilization plant, and step aeration plants.

Conventional Plant--A conventional plant consists of a series of rectangular tanks with a length to width ratio of 5 to 50. Tanks hydraulics are plug flow. In the system, sludge from the aeration tank is returned to be mixed with influent coming into the tank. When the safety factor for a conventional plant is compared to that of a complete mix plant, it is determined that the contentional plant is more efficient. However, the conventional plant has the disadvantages that the carbonaceous oxygen demand is concentrated at the head of the tank, making it more difficult to supply enough air in the front of the tank to supply both carbonaceous and nitrogenous oxygen demands. Air diffusion systems for conventional plants must be specifically designed to handle this concentrated load, otherwise, the first portion of the tank is not available for nitrification and the nitrification volume is reduced.

Complete Mix Plants--A complete mix plant is designed so the effluent entering the plant's aeration tank is evenly distributed. This design provides uniformity of load to all points and elimination of the oxygen transfer problems of the conventional plant. Due to the increased amount of short circuiting of the influent to effluent, ammonia concentrations may be slightly higher than a conventional plant. Load characteristics are similar to a conventional plant. Temperature plays a critical roll in nitrification. At a minimum temperature of 10°C, the hydraulic retention is about 14 hours whereas at a 20°C, the retention time is decreased to 6 hours. Therefore, location and temperature of the plant must be determined before sizing.

Extended Aeration Plant--Extended aeration plants are similar to complete mix plants except that hydraulic retention times are 24 to 48 hours instead of 2 to 14 hours used in the complete mix plants. Endogenous respiration is maximized in this type of plant, consequently 25 to 35 day solids retention times are not uncommon. Because of the long aeration time, they do suffer heat losses because of this long net growth, extended aeration plants can be expected to nitrify unless the temperature is less than 10°C.

Contact Stabilization Plants--Contact stabilization plants are not well suited for complete nitrification because insufficient biological mass is present in the contact tank for complete nitrification of ammonia. Since the ammonia is not absorbed on the biological floc, ammonia will bleed through to the effluent. In contact stabilization, instead of mixing the influent wastewater with return sludge, the return sludge is reaerated in a sludge reaeration tank prior to mixing with the influent wastewater. Backmixing between the contact and sludge reaeration tanks is prevented by the use of weirs. BOD_5 removal, not nitrification, is fairly high in this system because the bulk of the organics in domestic wastewater are particulate and can be absorbed by biological solids for later oxidation in the sludge reaeration tank.

Step Aeration Plant--In the step aeration plant, the influent wastewater is introduced at several points along the aeration tank even though the return sludge is introduced at the head of the plant like a conventional plant. This influent distribution reduces the initial oxygen demand usually experienced in contact stabilization plants and is avoided in a step aeration plant because of the longer contact time. In any case, some NH_4 bleed through will occur from short circuiting of the influent to the effluent and insufficient contact time for complete organic nitrogen hydrolysis and oxidation of ammonia.

Advantages and Disadvantages--The combined systems single stage is more compact than the separate stage system, but at the same time the chances for toxic upset are greater. Suspended growth systems (both combined and separate stages) have the advantage over the attached growth system in that effluent ammonia concentration are lower, but this is overcome in attached growth systems by treating the effluent by breakpoint chlorination. Also, in attached growth systems, stability is not linked to a secondary clarifier as in the suspended growth system. Separate stage attached growth systems are less sensitive to cold weather operations than the combined attached growth systems.

Denitrification--Biological denitrification is used to remove nitrogen from wastewater when it is in the nitrite or nitrate form. In the process, ammonia-nitrogen in wastewater is first converted to nitrite and subsequently nitrate. The nitrate nitrogen is then converted to a gaseous nitrogen, primarily nitrogen gas. This gaseous form of nitrogen is not objectional. This is accomplished through suspended and attached growth systems using methanol as the carbon source and combined carbon oxidation/nitrification/denitrification systems, using wastewater as an endogenous carbon source.

Pseudomonas, Micrococcus, Achromobacter, and Bacillus are principal nitrate-reducing bacteria that accomplish denitrification. These bacteria accomplish nitrate reduction by nitrate dissimilation whereby nitrate or nitrite replace oxygen in the respiratory processes of the organism under anoxic conditions. Using methanol as the carbon source, the energy reaction is represented as follows:

Step 1: Nitrate to Nitrite

$$6NO_3 + 2CH_3OH \rightarrow 6\ NO_2 + 2CO_2 + 4H_2O$$

Step 2: Nitrite to Nitrogen Gas

$$6NO_3 + 3CH_3OH \rightarrow 3\ N_2 + 3CO_2 + 3H_2O + 6OH + 3O_2$$

Overall energy reaction:

$$6NO_2 + 5CH_3OH + 3O_2 \rightarrow 3\ N_2 + 5CO_2 + 7H_2O + 6OH$$

A typical synthesis reaction is:

$$3NO_3 + 14CH_3OH + CO_2 + 3H \rightarrow 3C_5H_7O_2N + 19H_2O$$

Overall nitrate removal:

$$NO_3 + 1.08CH_3OH + H \rightarrow 0.065C_5H_7O_2N + 0.47N_2 + 0.76CO_2 + 2.44\ H_2O$$

where nitrate, nitrite and dissolved oxygen are present, the methanol requirement (C_n) can be computed using the following:

$$C_n = 2.47\ (NO_3\text{-}N) + 1.53\ (NO_2\text{-}N) + 0.87\ (DO)$$

typically, methanol feed ratios require three parts methanol per one part nitrogen by weight.

Suspended Growth Systems

Suspended-growth denitrification is usually carried out in plug-flow type activated-sludge systems. The anaerobic bacteria obtains energy for growth from the conversion of nitrate or nitrite to nitrogen gas but requires an external source of carbon for cell syntheses. Methanol is commonly used as this carbon source. In denitrification, it is nitrite or nitrate that is the pollutant to be removed, where as in conventional treatment, carbon is the pollutant. The methanol added is a carbon source but should not be added in excess, if left in the effluent, the methanol will exert a BOD_5. Because the nitrogen gas formed in the denitrification reaction hinders the settling of the mixed liquors, a nitrogen-gas stripping reactor precedes the denitrification clarifier. Removal of residual methanol induced BOD_5 is an added benefit of the stripping tank.

Attached Growth Systems

Denitrification in attached growth systems, are accomplished by a variety of column configurations. Column systems are nitrogen gas packed beds, submerged high porosity packed beds, and submerged high porosity fluidized bed systems. Since there is a variety of media used to measure reactions, general design criteria for these systems is not readily available. In order to size the reactor volume, one approach is to estimate the level of biomass on the media surface and then use measured reaction rates per unit of biomass to obtain the nitrogen removal capacity of the column. This is best done through pilot plant experiments. Based on the measured rates, surface nitrate removal rates are classified as high surface removal rates, reflecting the formation of extensive biological film; and low surface removal rates, reflecting minimal surface film development.

Nitrogen Gas Filled Packed Bed Reactor--The nitrogen gas filled packed bed reactor is covered and filled with nitrogen gas. This eliminates the necessity of having to submerge the medium to maintain anoxic conditions. The column media consists of plastic (corrugated sheet modules) similar to those used in trickling filters.

Aquatics

The use of aquatic plants as nutrient removal agents, is not a new concept. The water hyacinth (Eiclornia crassipes) is the most researched plant in this area to date. The water hyacinth is a floating plant that covers vast areas of water surface in the southern United States. The plant interferes with navigation, causes flood control problems, and restricts recreational activities such as fishing, boating, and water skiing. The water hyacinth is a native of South America and was introduced to the United States in the late 1800s. Any constructive use of this plant is most welcomed.

In 1968, 61% removal of PO_4-P was accomplished by growing water hyacinths after a five day detention time. However, most tests to date have shown that after 25 to 30 days of continuous operations, phosphate removal efficiency declines until only 5 to 8% removal efficiency is observed. Also, hyacinth removal efficiency is much less during colder months. Research done at Gainesville, Florida, found that the nutrient removal capacity of water hyacinths was directly related to pond surface area. In order to remove 44% of the phosphorus, a one million gallon pond with 5.1 acres of water hyacinths was needed. For the small ponds used in the Gainesville test, influent values ranged from 3.37 to 3.44 mg/l with effluent values of 1.82 to 1.86 mg/l. Effluent had a four-day detention time and the pond a depth of one foot. Tests were also done with a deeper pond and it was found the shallower pond, had better the PO_4-P removal efficiency. Nutrient uptake by the hyacinths was good during the area growth phase and vertical growth phase, but if the hyacinths growth was not limited, lesser efficiencies in PO_4-P were noted; however, some phosphorus removal is obtained.

Concluding Remarks

Biological treatment systems contain living organisms and require carbon, nutrients, and water as a vital part of the process. Such systems are affected by temperature, light, movement, and chemical conditions of pH, oxygen, salinity, and metals and the interrelations of the organisms present. Microorganisms perform best under steady state environment. Environmental changes if at suitable rates will allow the biomass to acclimate.

Biotechnology research for the future is striving to develop new strains of microorganisms and processes for destruction of organic toxics and hazardous materials heretofore immune to treatment and to shorten process and retention time. The need is present in the market place and biotechnology will respond.

2 BIOLOGICAL DEGRADATION OF HAZARDOUS WASTES

Industrial waste problems, especially those involving hazardous waste, are an ongoing problem. They have a high profile, and as such are reported regularly in the media. Examples of past mishaps are Love Canal, New York; Times Beach, Missouri; and Serveso, Italy. In our modern day world, with our use of plastics, cars, leather, and pesticides, it is now accepted that they have "side effects." Within the last number of years, there has been a move to clean these problems. Presented here is a review of current studies concerning biological treatment of hazardous waste, along with non-biological strategies.

Biological treatment is an expanding field which has shown success in remediation of hazardous waste problems. It is not the only treatment technology available and does not have to be used in isolation, nor is it without limitations. The chapter begins with an introduction, explaining some of the problems with hazardous waste, ways to reduce it, some past legislation, and facts about hazardous waste. Then there is a brief section reviewing current abiotic treatment methods available, including some of their advantages and limitations. Discussed are some of the types of biological treatment techniques, (e.g., land composting, activated sludge processes, filters), before assessing what we know about the microbiological basis for biological degradation. Also addressed is how we test if a compound is susceptible or not, the genetical basis of degradation, and the terms "mineralization" and "recalcitrance." There follows work done in degradation methods using aerobic bacteria in pilot plant demonstrations or in-situ studies. These are the success stories that have originated in laboratory research and bench scale demonstrations. A review of current laboratory work follows. The work discussed to date has been aerobic but there are anaerobic bacteria that degrade many compounds with certain advantages over aerobic degradation.

They have a different type of metabolism. This will be discussed prior to the laboratory work. Included in this section and not in the pilot studies/in-situ section is work on 1,1,1-trichloroethane, which is an in-situ study. It is placed here because it is an example of anaerobic degradation. The work with fungi has so far been ignored. Although there has been a great deal of information about fungal degradation in the past years, the majority of published works dealt with bacteria.

INTRODUCTION

Hazardous waste does not consist of one chemical or attribute; the definition of it is a wide one. The Environmental Protection Agency (EPA) stated in its report to Congress that:[1] The term "hazardous waste" means any waste or combination of wastes which pose a substantial present or potential hazard to human health or living organisms because such wastes are lethal, non-degradable, persistent in nature, biologically magnified, or otherwise cause or tend to cause detrimental cumulative effects. General categories of hazardous waste are toxic chemical, flammable, radioactive, explosive, and biological. These wastes can take the form of solids, sludges, liquids, or gases. So, the range of hazardous wastes, and their effects is large and diverse.

Attributes of hazardous wastes are: irritation, corrosivity, and bioconcentration. The observable effects are a function of exposure, the longer a person is exposed, the more serious the effects.

Nearly all industry produces some waste and since no one is likely to independently shut down an industry that makes the waste, we need some guidelines and regulations. On the federal level, industrial waste was originally regulated by the Clean Water Act of 1977 and the Clean Air Act of 1963 and subsequent amendments; hazardous wastes received special emphasis in the Resource Conservation and Recovery Act of 1976 (RCRA), the Toxic Substances Control Act of 1976 (TOSCA), and the Comprehensive Environmental Response, Compensation and Liability Act of 1980 (CERCLA) and subsequent amendments.

Through CERCLA, the Superfund aid was initiated and it was seen as an assault on growing hazardous waste problems, but to date, only 12-15 sites have been detoxified and restored. Most other sites have been limited to containment. The EPA estimates that there are some 2000 hazardous waste sites, but more alarmingly the Office of Technical

Assessment believes that 10,000 sites will eventually need detoxification. Meanwhile the U.S. industry continues to produce new wastes at the rate of 260 million metric tons per year.[2]

The EPA states that the preference for waste management priorities should first be via waste minimization techniques and then only after this is to focus on treatment.[3] There has been emphasis on the treatment abandoned hazardous waste dumps, preventing the illegal dumping of wastes, and regulating industrial discharges and dumping. But there has been a lack of strategies to deal with both the toxicity and volume of industrial waste at the source. Valentino and Walmet[4] discuss this process problem and discuss strategies to deal with it. They list as five most common strategies from generally least costly to most costly:

- Reduction in the overall amount of waste that is produced. The materials used in the front end of the process can be reduced by good housekeeping policies. There are numerous examples of this strategy. For example, an electronics firm has reduced the overall amount of solvent needed to degrease metal parts, and an electroplating shop has, instead of rinsing all parts with fresh water, made minor changes that allow the plater to "cascade" rinse water i.e., use water from the third rinse to perform the second rinse and so forth, thus cutting the total quantity of water requiring treatment by up to 90 percent.
- Reuse of the processing chemicals or solvents with little or no treatment, sometimes within the same company. Often, however a waste that cannot be used by one company can be reused somewhere else. For example, a waste alkali can be used to neutralize an acid stream at another facility. Cement dust captured in the pollution control equipment of a cement kiln is very useful in final drying and in stabilizing metallic sludges to prevent leaching after land disposal.
- Recycling the processing chemicals through a treatment process. The characteristics of the useful end product may be altered, but recycling may produce useful materials and energy. An example is the reclamation of titanium from utility-plant precipitator ash or precious metal from plating wastes.

- Research, development, and demonstration (RD&D) of new waste-reduction technologies. RD&D can range from basic research, to demonstrating the feasibility of tailoring an existing technology to the process of a particular business. An example is the development of a system for the anaerobic digestion of cheese-whey, a by-product of the cheese manufacturing process. While it is not a hazardous waste, cheese whey has an oxygen demand 350 times stronger than normal sanitary domestic sewage, and it creates serious disposal problems. In the past, many cheese-manufacturing plants have either spread the whey on farmland (which results in odor and run-off into surface water) or sent the whey down the sewers into a municipal sewage treatment works, which overloaded the treatment capacity of the system. The development of an anaerobic treatment system for cheese whey in which certain bacteria are used to convert waste material into a mixture of harmless gases, including methane is reducing the load of the systems and producing sizable quantities of biogas, which can be used as fuel.
- Replacement of the entire industrial process. This is the most costly strategy and usually beyond the capabilities of all but the most prosperous and willing industries. An example of replacement, which has applied to industrial as well as residential practices, is the wholesale reduction of oil-based paints in favor for water-based paints. Another example that illustrates a costly alternative is heat-treating certain metal parts to provide suitable hardness. Small items were previously hardened in molten cyanide baths. Because cyanide is toxic, many cyanide baths have been replaced by controlled atmosphere furnaces that are much more complex and costly to operate.

Legislation for the process problem would be difficult since each industry is different, with different starting chemicals, techniques, and products. Another problem with legislation is that waste problems vary on a regional basis and with locale. The waste problem in central New Jersey will not have the same characteristics as one in California.

Weather patterns differ greatly across the United States. Leaching of landfills is much less of a problem in arid regions than it is in the rainy areas. Solar evaporation may be an effective waste-concentration option in New Mexico but not in New York. Similarly, certain industries that are regional in nature have specialized waste treatment problems. For example, the petrochemical industry is concentrated on the Gulf Coast. Mining and ore refining tend to occur in only a few major locations. Steel has historically been heavily concentrated in Pennsylvania and Ohio. New York has unusually large concentrations of many small electroplaters in the New York City area, leather tanneries in two Mohawk River Valley counties, and a large number of diverse, old manufacturing operations in the Buffalo area and Elmira areas.

Most of the production of waste in the U.S. is from industry which is responsible for 69%. Municipal output accounts for 35%.[5]

The bulk of potentially hazardous waste is generated from four industries: primary metals; organic chemicals and explosives; electroplating and metal finishing; and inorganic chemicals. The hazardous waste management industry often deals with input materials in the five groupings of: metals/metal finishing; paint/solvent/coatings; organics; petroleum; and inorganics.

Some estimates show that 71% of the waste is produced from the chemical and petroleum industries, 22% from metal and related industries, and 7% from other industries.[5]

As stated before, waste production is regional in nature. Of the hazardous waste produced in the U.S. some 60% of it is from the following states: New Jersey, Ohio, Illinois, California, Pennsylvania, Texas, New York, Michigan, Tennessee, and Indiana.

Legislation as a tool in itself is not enough to deal with the problem; economically viable alternatives are necessary. Different states show different types of waste management, but land treatment is ubiquitous in treatment technologies. In the past the land took much of the brunt with landfills and surface impoundments, with deep injection into the groundwater system being the next biggest category, and other forms being responsible for only a small percentage. With the rising cost of land treatment, depletion of space, and concern over contamination of our groundwater, other technologies will have to take on a more significant role in treating hazardous wastes.

ABIOTIC TREATMENT TECHNIQUES

Wastewater Treatment

Selection of a treatment process is dependent on the nature of the wastewater and the quality of the effluent desired. Hazardous components of the wastewater may be either separated or converted to non-hazardous forms in order to permit the disposal of the wastewater effluent by conventional methods. Conversion processes can be done in one step or in multiple steps. The hazardous components which are separated from the wastewater must be disposed of. This may take additional steps, e.g., sludge dewatering.

Liquids-Solids Separation

Separation of suspended matter from wastewater can be accomplished by a number of different processes. Large heavy solids are easier to remove than finely divided light solids.

Screening devices are used to remove large pieces of solid matter that would interfere with subsequent processing operations or would cause damage to equipment such as pumps. Coarse screening devices may consist of parallel bars, rods or wires, perforated plates, gratings, or wire mesh.

Gravity sedimentation--This process involves the containment of wastewater for a sufficient period of time to allow some or all of the suspended materials to either settle out or float to the surface of the wastewater. In its simplest form as a batch process, a given volume of wastewater is transferred to a vessel and held there until nearly all the settleable and floatable matter separates. The floating matter can be skimmed off and the wastewater decanted for discharge or further treatment. Sludge may be allowed to collect until several batches of wastewater have been processed. Then it is removed. The vessel may have a conical bottom so that the sludge can be removed via a valve. Large settling ponds may be constructed which are drained periodically to permit sludge removal. Solids-contact or sludge-blanket clarifiers are useful where sludges are flocculent and of low density. They are designed with large mixing and reaction zones that coupled with the sludge blanket account for greater efficiency in solids removal. Gravity sedimentation works like the clarifiers but more time is taken for the settling.

Dissolved-Air Flotation is useful for suspended matter that does not sink or float in a reasonable period of time. Separation is brought about by the introduction of finely divided gas bubbles which become attached to the particulate matter, causing it to float to the surface where it is removed by skimming. Introduction of the gas bubbles is usually accomplished by reducing the pressure of the wastewater causing dissolved gases to be released. This is commonly used to separate greasy or oily matter from industrial wastes. Granular-Media Filters or deep bed filtration is a polishing step that removes small amounts of suspended solids and produces a highly clarified water. Chemical coagulation and sedimentation typically precede this stage. Graded sand and pulverized coal are commonly used in the filter beds. Conventional operation is usually by downflow. The ability of the granular media filter beds to produce a clear effluent results from the straining action and adhesion, which removes particles finer than the pore space.

Surface Filters make use of a fine medium such as a cloth or close mesh screen. In the rotary vacuum filter, the medium is in the form of a continuous belt and it rotates over a perforated drum that is partially submerged in the slurry to be filtered. Water is pulled through the filter cake that forms on the belt to the inside of the drum, where it is transferred to the vacuum system.

Centrifugation is a useful alternative to filtration for sticky sludges that do not dewater rapidly on a filter. They operate by a rapid rotation of a liquid suspension, which induces a much greater force than gravity to hasten the separation of the suspended matter.

Chemical Treatment

Chemical treatment is a widely used process for the destruction or separation of hazardous constituents in wastewater. This can be done by neutralization of acidic or alkaline wastewater until a suitable pH is obtained. Precipitation/Coagulation/Flocculation is used for the removal of heavy metals. Precipitation refers to the formation of a solid phase, coagulation is where the containment is trapped by the formation of a precipitate, and flocculation is the agglomeration of a coagulating chemical.

Oxidation-Reduction or the redox processes are used for converting toxic pollutants to harmless or less toxic materials that are more easily removed. These processes involve the addition of chemical reagents to wastewaters, causing changes in the oxidation states of substances both in the reagents and in the wastewaters. In order for one substance to be oxidized, another must be reduced. Ozone is a powerful oxidizing agent that is usually oxygen at temperatures and pressures up to 350°C and 180 atmospheres, respectively, to treat organic wastes.

Ion exchange involves a change in the chemical form of a compound; the exchange of ions in solution with other ions held by mixed anionic or cationic groups or charges. Typically, a waste solution is percolated through a granular bed of the ion exchanger, where certain ions in solution are replaced by ions contained in the ion exchanger. If the exchange involves cations, the exchanger is called a cation exchanger and correspondingly anion exchanger is one that involves anion exchange.

Physical Methods

There are several methods used for separating pollutants from wastewater: activated carbon, steam stripping, evaporation, reverse osmosis, and solvent extraction. The chemical and physical characteristics of the pollutant are important in the selection of the physical removal method. Steam stripping is effective for substances that have an appreciable vapor pressure at the boiling point of water, whereas evaporation is effective for those chemicals that will not volatilize. Soluble, small organic molecules are adsorbed by activated carbon; large ions are separated by reverse osmosis.

Activated Carbon Adsorption--Here the inorganic and organic chemicals are adsorbed onto activated carbon. Usually hydrophobic chemicals are more likely to be removed. The degree of adsorption is linked to the molecular weight, methanol-water coefficient, or solubility (these are also linked to the recalcitrance and/or toxicity). The smaller the size of the grain, the more surface area is available and so equilibrium is reached quicker with powdered activated carbon compared to the granular form. But then the powdered form needs more pumping to get the wastewater through and hence the costs are increased. There are two principle systems, one is downward flow through the bed (pressure or gravity flow) and the other is upflow through a packed or expanded bed. Activated carbon adsorption is applicable to the treatment of dilute aqueous wastes, but they should be treated to remove suspended solids, oil, and grease. Temperature and pH are also important for the different compounds to be treated. The carbon is either disposed of or regenerated. Carbon has also been added directly to biological treatment effluent in a contacting basin. The advantages of this are that the sludge toxicity is reduced by selectively removing the toxic organics from solution and that the carbon adsorption capacity is extended by bioregeneration of the "biocompatible" species adsorbed on the surface.[6] For aqueous solvent waste containing contaminants in concentrations up to 10,000 mg/l, the activated sludge process has been proposed as a potential applicable treatment. However these concentrations may be toxic to the sludge or they may be easily stripped to the atmosphere, thereby creating another hazard. The sludge may also contain recalcitrant waste, due to adsorption of the contaminants and be difficult to dispose of.[7]

Evaporation is the process that heats the liquid, venting the vapors to the atmosphere and concentrating the pollutants into a slurry.

Reverse osmosis--Osmosis is the process where a solvent (e.g., water) moves from an area of low concentration to high across a semipermeable membrane which does not allow the dissolved solids to pass. In reverse osmosis, a pressure greater than the osmotic pressure is applied so the flow is reversed. Pure water will then flow through the membrane from the concentrated solution.

Solvent extraction is a process whereby a dissolved or adsorbed substance is transferred from a liquid or solid phase to a solvent that preferentially dissolves that substance. For the process to be effective, the extracting solvent must be immiscible in the liquid and differ in density so that gravity separation is possible and there is minimal contamination of the raffinate with the solvent. The hydrophobic solutes are more likely to be extracted. Solvent extraction can be performed as a batch process, or by the contact of the solvent with the feed in staged or continuous equipment.

Steam stripping is where water vapor at elevated temperatures is used to remove volatile components of a liquid. Countercurrent flow is generally used to promote gas-liquid contact, thus allowing soluble gaseous organics from the liquid waste to be continuously exchanged with molecules within the stripping gas. Again, this is useful only for waste with low water solubilities.

Incineration

Incineration is a high temperature oxidation process that converts the principal elements (carbon, hydrogen, and oxygen) in most organic compounds to carbon dioxide and water. Given with the problems of disposing on land, incineration may take on a lead role in waste treatment. However, it is not without its problems. There is fear among the general public about the nature of the stack emissions but it is an efficient method. The destruction of the molecular structure usually eliminates the toxicity of the chemical. But the existence of other elements in a waste may result in the production of particulate pollutants that require removal in off-gas treatment systems. There are several types of incinerators available.

The **liquid injection incinerators** operate by spraying the combustible waste mix with air into a chamber where flame oxidation takes place. The purpose of spraying is to atomize the waste into small droplets which present a large surface area for rapid heat transfer, thereby increasing the rate of vaporization and mixing with air to promote combustion. Air is supplied to provide the necessary mixing and turbulence. These incinerators are widely used for destruction of liquid organic wastes.

Rotary-kiln incinerators are designed to process solids and tars that cannot be processed in the liquid incinerator. The rotary-kiln is a cylindrical shell lined with refractory material that is horizontally mounted at a slight incline. It is rotated from 5 to 25 times at high temperatures, 1500 to 3000°F with excess air, the residence time varied depending on the nature of the waste. The rotation causes a tumbling action that mixes the waste with air. The primary function is to convert, through partial burning and volatization, solid wastes to gases and ash/residue. If the ash is free of dangerous levels of hazardous wastes, it is put in a landfill.

Fume incineration--Large quantities of organic vapor fumes are produced by many industries, including fat rendering, metal painting and varnishing, and various types of printing. These vapors are generally mixes of hydrocarbons, alcohols, and acetates. The mixes may not be acutely toxic but they do cause odor problems. An integrated heat-recovery system demonstration at Case-Hoyt Company, near Rochester, captures waste heat and simultaneously reduces plant air emissions by oxidizing solvents into harmless gases as the heat is recovered.

The **multiple hearth incinerator** is used for wastes that are difficult to burn or that contain valuable metals that can be recovered. It consists of a refractory-lined circular steel shell, with refractory hearths located one above the other. Solid waste or partially dewatered sludge is fed to the top of the unit where a rotating plow rake plows it across the hearth to dropholes. The uncombusted material falls to the next hearth and the process is repeated until the combustion is complete.

Fluidized-bed incinerators (FBI) are applicable to the destruction of halogenated organic waste streams. This type of incinerator consists of a vessel in which inert granular particles are fluidized by a low velocity air stream which is passed through a distributor plate below the bed. An FBI consists of a windbox (through which combustion air is introduced to the reactor), and a reactor zone (containing a bed of sand, waste injection, and removal ports). Temperatures are in the range of 1300 to 2100°F, gas residence times usually a few seconds. They have been used to treat municipal sewage sludge, low quality fuels, pulp and paper effluents, food processing waste, refinery waste, radioactive waste, and miscellaneous chemical waste.[9]

A **molten salt incinerator** uses a molten salt such as a sodium carbonate as a heat transfer and reaction medium. In the process, waste material along with air is added below the surface of the bed so that any gases formed during combustion are forced to pass through the melt. Reaction temperatures in the bed range from 1500 to 2000°F and residence times are less than a second. Any acidic gases formed are neutralized by the alkalinity of the bed. This can change the fluidity of the bed so it needs replacement frequently.

Plasma arc incineration is based on the concept of reducing or pyrolyzing waste molecules to the atomic state using a thermal plasma field. They system uses very high energy at temperatures near to 10,000°C to break bonds of hazardous waste chemical molecules down to the atomic state. An electrode assembly ionizes air molecules which create a plasma field. Hazardous waste mixtures interact with the field, forming simple molecules such as carbon dioxide, hydrogen, hydrogen chloride, and other minor matrix compounds such as acetylene and ethene. Westinghouse Electric has a mobile plasma arc unit called Pyroplasma that reportedly treats liquid wastes at the rate of a 3 gal/min.[10] The high temperatures decomposes PCB and other wastes in an oxygen-deficient atmosphere. Hydrogen chloride is treated with sodium to form water and salt.

Lime or cement kiln incineration--A cement kiln is basically a large rotary kiln in which raw materials are fed countercurrent to combustion gas flow. The wet process kilns uses a 30% water slurry feed and are the most suitable for hazardous waste destruction. The products formed are alkaline and so act as a scrubber, removing acid gases formed during combustion. This system operates at 2800°F resulting in very efficient removal of wastes. A project[10] is in operation at Blue Circle Atlantic Inc., a cement company in Ravena where it is evaluating the incorporation of hazardous wastes into an existing cement-kiln operation. The facility is not yet in operation but should be able to handle large amounts of waste and alleviate some problems locally.

Wet Air Oxidation

Wet air oxidation involves the aqueous phase oxidation of organic materials at high temperature and pressure. A major advantage over other incineration methods is that the water in the waste stream is kept in the liquid state. Water is pumped into the reactor along with oxygen which is heated by the hot effluent. Two types of reactors are used, a bubble tower reactor and a stirred tank cascade reactor. This process is good for wastes that are too dilute to incinerate but too toxic for biological methods. The products are usually acetic acid and carbon dioxide.

Solidification Techniques

There are several innovative non-thermal processes that have been developed under the SITE program that immobilize wastes by vitrification or other types of solidification.[10] The SITE program is the Superfund Innovative Technology Evaluation, a $20 million/year program which has been developed to encourage the private development and demonstration of new technologies for cleaning up hazardous wastes. For example, the researchers at Battelle Pacific Northwest Laboratories have developed an in-situ vitrification process (which was originally designed for the containment of nuclear wastes) in which electrodes are sunk into a contaminated area and attached to a diesel powered generator. The current produced temperatures of about 3600°F which is much higher than the fusion temperature of soil. An exhaust hood is placed over the site to collect and treat any combustion products. The result is a massive glass-like product consisting of completely immobilized organics, inorganics, steel drums, and other components that are essentially locked up and inert. The time taken to complete the process depends on the electrode depth and frequency. Another solidification process uses a reagent called Urrichem that immobilizes slurried hazardous components. The contaminated soil is excavated and intimately mixed with the Urrichem off-site. After blending, the slurry is pumped out of the mixer and hardens into a concrete like mass within 24 hours. Chemfix is a process developed by Chemfix Technologies (Metaire, LA). In this technique, a proprietary blend of soluble silicates and additives is used to convert high molecular weight organic and inorganic slurries into a cross-linked, clay like matrix.

Alternative methods have also included landfills, deep well injection disposal, and ocean dumping. Landfills were developed because it was believed that by placing waste in designated ground areas, there would be a natural decomposition over time. Unfortunately the water table rises in a landfill and this mounding effect means that we get a leaching of the water containing toxics since the water flow is always from areas of high to low head, i.e., there is a gradient set up that favors water movement originating in the landfill and away from it thus contaminating our groundwater supplies. This is a special problem for an area like Long Island, New York, that relies on groundwater and not surface water for their domestic/industrial supplies. Deep injection wells have similar problems. When the aquifers in which the deep well sits are pumped, the contaminants are drawn into the well. Ocean dumping has upset the delicate ecosystem balance in some areas. Thus there is a need for effective techniques that industry can safely use to try and combat the nation's hazardous waste problems. Biological treatment may be such.

BIOLOGICAL CONTROL METHODS

Biological treatment of hazardous wastes is becoming a realistic option in the treatment strategies for hazardous wastes. As we research more about specific bacteria and fungi that do the work and know more about what conditions favor degradation, we are finding more uses for them. A major advantage is that the cost is much less than incineration. A limitation is the length of time involved, since biological treatment can take months. There are different mechanisms that we can use to degrade a compound. Selection is based on the type of pollutant, type of degradation (i.e., aerobic or anaerobic), and various other parameters.

There are suspended growth systems such as activated sludge, aerated lagoons, and anaerobic digesters where the microorganisms are suspended; and fixed film processes, which include filters and rotating biological contactors where the microorganisms grow on a fixed surface. Since toxin degradation is a slow process, the fixed film processes may be best since the microbial retention is very high.[11] In aerobic degradation, the major portion of the cost is the aeration. Wastes in any form--solid, aqueous, and gaseous--can be treated biologically. The treatment can be done several ways: by applying the waste to the soil, by composting, using a hybrid liquid/solid treatment technique, treating the wastes in-situ, using soil filter gases, and with wastewater treatment systems.[12] Care must be taken in assessing biological treatment since non-degradable pesticides and other organics may be unaffected by the treatment or toxic intermediates may be formed and therefore a finishing process such as activated carbon may be needed.

Land Treatment

Land treatment is the biological option most widely used to treat hazardous wastes. Many of the industrial waste treatment sites using biological treatment are at petroleum refineries.

The steps in the process typically are: spread the waste on the land, let it dry, till the soil to mix the waste in, control moisture, and if necessary add nutrients so that the waste-destroying bacteria will grow.

In a land treatment system, bottom liners under the treatment zone are generally not needed. Physical, chemical, and biological processes in the soil are relied on to degrade, immobilize, or transform the wastes to the environmentally acceptable forms. Most pollutants are captured and transformed in the top 6-12 in. in the soil, though the treatment zone may extend down 5 ft.

Land treatment is suitable with wastes such as petroleum refinery sludges, creosote sludges and wastewaters, and processing sludges from wood, paper, and textile manufacture. It is also widely used for municipal sludges and wastewaters, and food processing sludges and wastewaters. Oil, metals, and other constituents of environmental concern are successfully controlled by land treatment.

While most land treatment sites are at operating facilities, at least one is at an abandoned wood preserving facility now being remediated under Superfund or CERCLA.[12] At this Minnesota site a treatment pilot test was run for a year proving that the creosote constituents could be degraded by acclimated soil microorganisms.

To scale up to full operation, about 12,000 cu. yd. of contaminated soils and sludges were removed from the impoundment and temporarily stored. A lined treatment area was then constructed over the impoundment, and the stored wastes were put back in place.

The process has been encouragingly effective, as shown by polycyclic aromatic hydrocarbon removals of 90% after 193 days of treatment. Total cost, including excavation of the waste and construction and operation of the facility, was approximately $80/cu. yd. This cost is much less than incineration, the only other in-situ treatment technique that completely eliminates the hazardous compound.

Costs for unlined land treatment systems are a little cheaper but since we need to protect the groundwater system, extensive site testing is necessary to obtain federal RCRA permits.

As a result of the Hazardous and Solid Waste Amendments of 1984 and the current land disposal restrictions rule, EPA is in the process of demonstrating achievable treatment techniques to be used as an alternative to the land disposal of hazardous wastes.[13] Research has been done by waste generators that also treat the waste, commercial facilities, and EPA in-house research.

Composting

Composting uses less space than land treatment and it controls gases and leachate. In composting, piles of the waste 3-6 ft. high are treated. Aeration is provided either by turning the piles mechanically or through a forced aeration system. Bulking agents such as wood chips are sometimes added to facilitate mixing and oxygen transfer. These bulking agents are typically screened out at completion of composting, and mixed in with the next batch.

Composting has been used mostly in treating municipal sewage sludge, but has also been used to treat several industrial solid wastes. These include industrial wastewater treatment sludge, food processing wastes, and some industrial wastes containing low levels of pesticides.

A flow diagram for a biochemical compost bed system is shown in Figure 1.

A pilot scale bacteria-based composting system has successfully been used for treating diesel-fuel contaminated soils.[14] Aeration is provided by an air compressor that draws a vacuum at the bottom of the pile and returns the air at the top. Additional air is added as necessary. Nutrients and water are provided through an irrigation system located at the top of the pile. The system is a fully enclosed vessel that is batch loaded. Studies to date have shown that the diesel contaminants have half-lives in the range of 30-60 days. Costs of a full-scale system are estimated at under $100/cu. yd. of soil, again significantly less than incineration.

Liquids/Solids Treatment Systems (LSTS)

These are hybrids, intermediate between land treatment and conventional water-suspended biological systems. The waste is not in a solid form as in land treatment, nor in water as in conventional municipal wastewater treatment, but half way between. The wastes are in a suspended solid, slurry or sludge form.

LSTS systems reduce the level of contaminants in a waste primarily by 1) dissolving the organics, 2) biodegrading the dissolved organics into less toxic and less environmentally significant forms, and 3) releasing the products as gases to the atmosphere. The key step, dissolving the wastes from solid to liquid, can result from microbial action and/or physical-chemical action.

There are two basic LSTS process types: single-reactor and two-stage. In either case it is vital to have sufficient mixing of the liquid/solid mixture to achieve effective mass transfer of both organics in the liquid phase, adequate oxygen transfer to the microorganisms, and to keep the suspension of solids in the solution.

The single-batch reactor is the simpler process and the most widely used. Here both the dissolving and biodegradation steps occur together in a single tank, pit, or lagoon. Air is introduced to provide mixing for the extraction process and oxygen for biological reasons. Aeration and mixing may be provided by submerged aerators, floating aerators, or compressor/sparger (lance) systems.

At completion of treatment, the aerators and mixing are stopped and the solids allowed to settle. The treated liquid is decanted off the top,

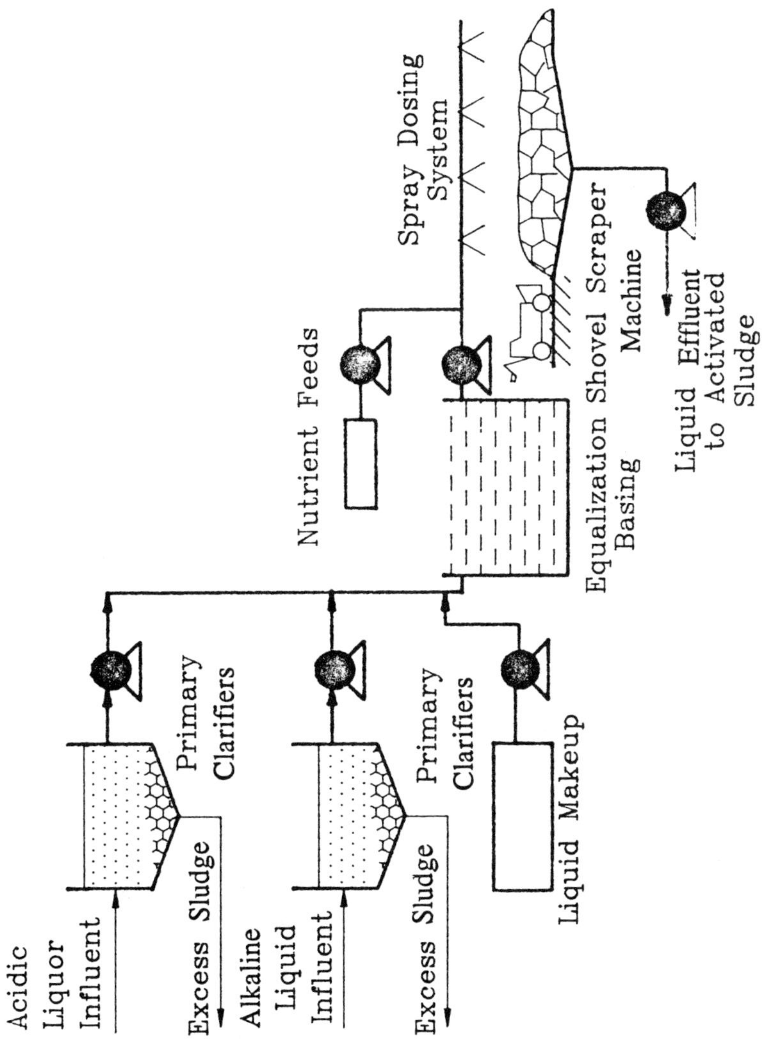

Figure 1. Biochemical compost bed system flow diagram.

while the treated solids are treated further, by conventional land treatment or by disposal on-site. If the process is done in a lagoon or surface impoundment, it may be possible to decant the liquid and leave the solids in place.

LSTS have been used in coal tar waste sites in Florida and Illinois.[14] In Florida, about 600,000 gal. of the wastes were found in a gas holder and in Illinois 2000 cu. yd. of material in an in-ground concrete structure. After 9 months, significant reduction of tar volume, phenol concentration and flash point were reported.

LSTS have the potential to achieve the treatment goals in a shorter time using less land area than other on-site biological management approaches, such as land treatment and composting. The process uses tank reactors, and thus may eliminate regulatory obstacles or concerns related to the possible land-treatment ban or to contaminant migration in the groundwater.

Soil Biofilters

Hazardous wastes in gaseous form can be biologically treated too. On-site biofilters can reduce nuisance odors, volatile organic compounds, and particulates in waste gases. The soil biofilter involves two mechanisms; mechanical filtration and treatment by bacteria on the soil particles.

A biofilter consists of a distributing pipe network underneath soil or gravel. The gas to be treated is pumped through the media and escapes through holes in the pipe wall. It rises through the media bed which is usually 3-10 ft. deep. Pollutants captured on the biofilter can be removed by filtration, physical and chemical absorption, chemical oxidation, and structural treatment.

The technique is not a new one, filters have been used for many years to remove odors from agricultural, food, and wastewater processing operations, and emissions from paper and chemical manufacturing plants. The soil filters have successfully removed low concentrations of H_2S, SO_2, NH_3, and NO_2 as well as particulates in these gases. Other potential applications include treatment of exhaust gases from air stripping units for wastewater and solvents, from mechanical technologies such as belt presses for oily sludges, from aerated waste treatment systems, and from in-situ soil stripping of volatile organic compounds (VOCs) using vacuum and forced aeration methods.

The S.C. Johnson Wax Co., in Racine, Wisconsin[14] has demonstrated removal of VOCs in a soil bed. This, in turn, has led to research on the biodegradation of chlorinated ethenes, such as those likely to be in exhaust gases under anaerobic conditions. The research demonstrated the aerobic degradation of TCE and other chemicals.

A soil biofilter is being planned for the treatment of VOCs being removed from contaminated soil and groundwater in New Jersey.[14] Natural gas will be introduced with the contaminated air stream to stimulate growth of the needed methane-oxidizing bacteria.

Wastewater Treatment

Biological treatment has been very successful in the removal of organic pollutants and colloidal organics from wastewater. Activated sludge, biological filters, aerated lagoons, oxidation ponds, and aerobic fermentation are some of the methods available for wastewater biodegradation. In removal of toxic waste, more care is needed since the bacteria are prone to destruction from shock loading or increases of toxic material fed in without allowing time for the population to grow large enough to deal with it.

Biodegradation occurs because bacteria are able to metabolize the organic matter via enzyme systems to yield carbon dioxide, water, and energy. The energy is used for synthesis, motility, and respiration.

With simple dissolved matter, it is taken into the cell and oxidized, but with more complex inorganics, enzymes are secreted extracellularly to hydrolyze the proteins and fats into a soluble form which can then be taken into the cell and oxidized. Hence the more complex matter takes longer to process.

Some organic compounds are "refractory," they cannot be oxidized while others are toxic to the bacteria at high concentrations.

The purpose of biodegradation is to convert the waste into the end products and material that will settle and can be removed as sediment. Again, biodegradation may not be one hundred percent, or toxic by-products may be formed. Further treatment by chemical methods or dilution may be needed to get the contaminant to a concentration prescribed as safe.

Nitrogen and phosphorus are essential in the oxidation process for the synthesis of new cells, and trace amounts of potassium and calcium are also required. The former are sometimes deficient so nitrogen is added in the form of ammoniacal nitrogen (nitrite and nitrate are not readily used by bacteria).

BOD or biochemical oxygen demand, measures the strength of the organics present and is defined as the amount of oxygen needed by the bacteria for oxidation. The more concentrated the organic material the higher the BOD. A BOD:N:P ratio of 100:5:1 is thought to be the optimum ratio of nutrients needed by bacteria.

Activated Sludge Process

This involves the generation of a suspended mass of bacteria in a reactor to degrade soluble and finely suspended organic compounds.

In this method the wastewater with its organic compounds is fed into the aeration tank. This is supplied with air and is vigorously mixed to allow maximum contact of bacteria and waste. The contents, referred to as MLSS (mixed liquor suspended solids) are then fed to a sedimentation tank where the treated solids settle to the bottom and the top liquid layer is treated and discharged. Part of the biological solids are recycled back to the aeration tank to maintain the correct mix; the remainder is waste.

This method is flexible and can be used on almost any type of biological waste. An industrial application has been demonstrated for phenol degradation using a petroleum refinery wastewater: there was an 85-90% removal of phenol and cyanide in the steel industry.[15]

Trickling Over Process

Here the wastewater is distributed by a flow distributor over a fixed bed of medium on which the bacteria grow forming a slime layer to which oxygen is supplied. The wastewater flows down over the slime layer which absorb organic materials and nutrients, releasing the oxidized end products to the drainage system underneath. Eventually some of the layer will detach with the wastewater, and then some additional separation is necessary.

Stabilization

This is a procedure where wastewater is stabilized by the actions of bacteria in shallow ponds. There are basically two types of ponds, ones where there is a natural supply of oxygen from algal photosynthesis (oxidation ponds) and mechanically supplied oxygen (aerated lagoons). The bacteria metabolize the wastes and the solids settle at the bottom as a sludge. Also there is anaerobic decomposition where the bacteria at the bottom will degrade the waste without oxygen's presence. Or there can be a lagoon that has both aerobic and anaerobic decomposition, with an interchange of products between the two layers of bacteria in a symbiotic relationship.

DETERMINATION OF BIOLOGICAL DEGRADABILITY

Some terms are important in understanding the fate of hazardous wastes in biological systems. They are *biodegradable*, *persistent*, *recalcitrant*, and *mineralization*. A compound that is biodegradable can be changed by the action of microorganisms to another compound. This does not necessarily mean that the product is less toxic than the parent compound, or that intermediate toxic compounds are formed that may inhibit the degraders, or that the product will not be toxic to the next degraders or to man himself. Park et al.[16] demonstrated this possibility in their study of bacterial degradation of 7,-12 dimethylbenzanthracene is non-acclimated soil. When this compound was degraded aerobically, the parent^{14} C DMBA had a half life of 17 days. There was a decrease in the parent compound accompanied by an increase in metabolites: 4-hydroxy, 5-hydroxy, and 10-hydroxy DMBA and 7,12 dihydro-12-methyl 7-methylene benz (a)anthracene. But there was some associated mutagenicity that was thought to be due to the polarity of the compounds, with those that are low and moderately polar having more mutagenic potential than the highly polar compounds. The pH of the soil also plays a role since at low pH fungi dominate the ecosystem and they are more likely to form epoxides (associated with mutagenicity) than bacteria which are dominant at higher pH. So there are some possible types of reaction that we do not want. We need information so that we can select favorable degradation reactions so as to reduce some types of toxic waste. Recalcitrant means that the compound cannot be biodegraded under any circumstances, i.e., it's the compound itself that is inherently resistant and not the treatment system that has failed to account for some vital fact. Persistence is a "conditional" property of the biodegradable compound in that it may be biodegraded if the correct circumstances favoring biodegradation are present.

Mineralization is the complete conversion of an organic compound to the end products of carbon dioxide and water. Primary biodegradation is the single transformation of a compound and partial biodegradation is somewhere between primary degradation and mineralization.

Basis for Biodegradation

Why are some compounds degraded by microorganisms? The degradation is done not as a favor to us but because the organisms gain energy needed for growth, repair, reproduction, and other biological functions needed for survival. Food sources contain oxygen in hydroxyl (OH) or carboxylic acid (COOH) groups. Oxidation reactions take place where electrons are transferred along an electron chain with compounds accepting and passing on electrons to a terminal electron acceptor. The coenzymes may be NADH and NAD which get reduced and then coupled with the electron transport chain to produce high energy bonds in ATP. Many compounds are biochemically inert such as alkanes, saturated ring structures, and unsubstituted benzene. They are devoid of oxygen and not subject to dehydrogenation reactions. The ability of bacteria to utilize these compounds lies in the fact that they can catalyze oxidation using oxygen. Other bacteria have enzymes which work without oxygen and coenzymes are needed.

Studies have been done that attempt to relate the physical and/or chemical properties of a chemical to its recalcitrance; however experience has shown that many chemicals which are thought not to be removed by abiotic mechanisms actually undergo transformations in model treatment systems.[17]

Genetics

Bacteria can only do things for which they have a genetic capability. They must be able to produce the right enzymes to do the job, and the right environment must exist for them to be able to produce the right enzymes. If a chemical is present in concentrations either too low or too high, then it may not be biodegraded: too low and the enzymes will not be induced, too high and the compound may be toxic. One reason why bacteria are robust at biodegradation is that they may still be able to degrade a different compound to the one that they normally utilize if the active site of the molecule has not been altered. The xenobiotic still needs to be able to induce the production enzymes or the reaction will stop. The bacteria will not be able to metabolize it and will die. This use of a different substrate is fortunate for us and has been termed "gratuitous" metabolism. Complicating biodegradation further is "cometabolism" where two substrates are needed i.e., one compound cannot fulfill all the bacterial needs and act as the sole carbon and energy source. A second compound is needed as growth substrate. Biological degradation treatment systems have to be designed to take care of another problem. Bacteria need a continuous carbon source for growth and yet if too much substrate is added, initially the bacteria are unable to metabolize it. They need a period of "acclimation" where they are growing and strengthening and perhaps even undergoing genetic changes. If the intermediate is toxic then the bacteria may be killed or the formation of the next strain that metabolizes it may be stopped so that it accumulates and destroys the system. Bacteria often work with a succession of strains so that complete mineralization needs more than one organism which does not alone have the required genetic capability. As will be discussed in the anaerobic section, mixed cultures can degrade better both qualitatively and quantitatively than single cultures. This may involve mutation. Vandenbergh et al.[18] took different strains of bacteria, found out what they could grow on and not; and then did transformation and conjunction experiments with them. They found that after plasmids had been introduced, strains that couldn't previously metabolize toluene, could now degrade it. They also isolated the plasmids necessary for activity. Their findings support the idea of evolution since the bacteria specific for the biodegradation of halaromatic compounds could not utilize toluene. There has been an evolution pathway involving mutations and transfer of genetic material between bacteria so that although at one time they could degrade toluene, they could now only degrade the products of it and not the parent compound.

A less desirable aspect of the ability of bacteria to degrade compounds has been reported by Reed et al.[19] who noted that some pesticides are degraded quickly by microorganisms before they have eliminated the pests. This phenomenon is called accelerated pesticide degradation. The study examined the response of microbial populations in soils after incorporation of the herbicides, alachlor, butylate, EPTC, carbofuran, cloethocarb, and isophenos. Enzyme assays for alkaline phosphomonoesterase, phosphodiesterase, and rhodanese were also studied. To test the sensitivity of selected microorganisms to the pesticides, formulations of the pesticides were applied to cultures on agar plates and zones of inhibition noted. Those bacteria that grew on pesticides were examined for specific enzymes. The activities varied but were greater for those in the field than control strains indicating that there had been a selection for increased activity in the presence of pesticides. Several isolates capable of growing on the carbamate insecticide, carbofuran, and the thiocarbamate herbicides, butylate and EPTC, also exhibited high rhodanese activity suggesting that this enzyme may be involved in the metabolism of these pesticides. These results suggest that the bacteria capable of developing high rhodanese activity in response to applications of thiocarbamate herbicides (e.g., butylate) may also be involved in the degradation of the chemically similar carbamate insecticides (e.g., carbofuran) without prior exposure to them. Again evolution appears to be the factor here; individual bacterium may acquire the ability to utilize a pesticide by the evolution of specific enzymes. Several organisms may have to coexist in a community to metabolize a pesticide. Assays for specific enzymatic activities of microorganisms shown to metabolize certain pesticide substrates indicate that cross-adaption of the microorganisms for degradation of chemically similar pesticides in the soil environment may exist. The growth pattern of microorganisms is conceptually illustrated in Figure 2. This is a classical viewpoint.

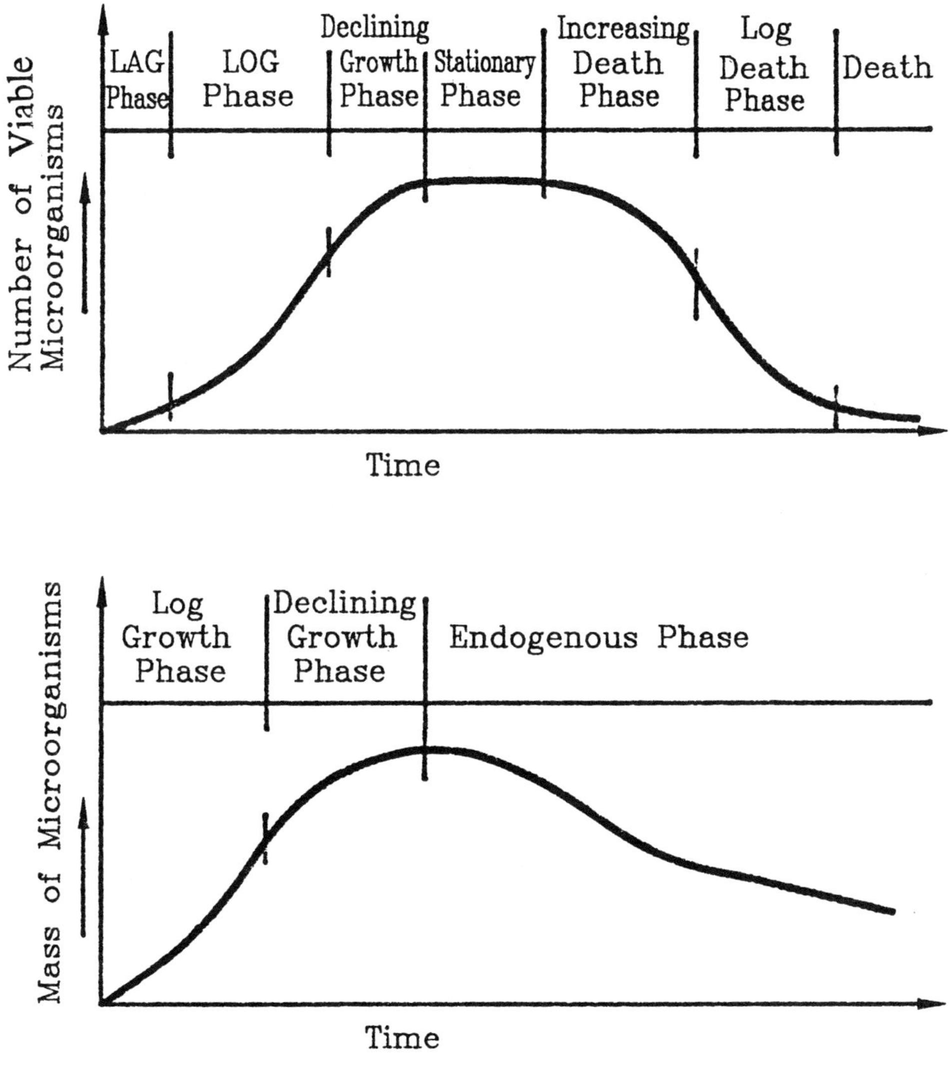

Figure 2. Growth pattern of microorganisms.

Testing for Recalcitrance

There have been several endeavors to find out which chemicals are recalcitrant and which are not and under what conditions. Obviously if we have a list of possibilities, then designing hazardous waste treatment becomes all the easier. However, there are problems and these will be discussed in this section. First we need an experimental technique that lets us test for recalcitrance. A "tiered" approach has been advocated by Grady.[17] The philosophy behind tiered testing is that tests are started that isolate those compounds which are readily biodegraded under less than optimum conditions and then give more and more favorable conditions up through the tiers until a compound can be classified as recalcitrant. This is a cost saving approach. Recalcitrance can only be deduced from the failure of biodegradation under the most favorable conditions. It cannot be proved.

Aerobic Tiered Testing

Tier 1 is a screening test where the test compound is the sole carbon and energy source to organisms that have not had prior exposure to it. Biodegradation is measured by dissolved organic carbon (DOC) removal, oxygen uptake, or carbon dioxide evolution. Cometabolism is precluded and so the genetic capability of the bacteria must be wide. The use of an unacclimated culture and the short time of the test means that induction must be the only mechanism responsible for the degradation. Only compounds readily degradable will give positive results. If biodegradation does take place, then we need to know if inhibition is important. Therefore growth of bacteria are tested with the substrate over a range of concentrations. For a non-biodegradable compound inhibition can be tested on another group of bacteria that are growing on a different substrate (i.e., are able to utilize that substrate unless they are inhibited by the test compound). *Tier 2* testing is acclimation and enrichment. Here a culture of bacteria is grown on more than one carbon source. The test compound is applied first in low concentrations and increased if there is evidence of biodegradation. Since there is more than one carbon source, DOC removal,. oxygen uptake, and carbon dioxide emission cannot be used. Here chemical analysis is needed specific for the test compound. *Tier 3* is the assessment of the degree of biodegradation. If we get a positive result from Tier 2, then we can go back and do a Tier 1 test with a mixed culture. If the results are positive, the test compound can be degraded by a culture as the sole carbon and energy source. If the results are negative at this point, it doesn't mean that the compound is recalcitrant, but that cometabolism is needed or that complete biodegradation is impossible. Following the fate of carbon can be done through radioactive labelling and seeing if the carbon ends up in new cells or metabolic intermediates which may then be mineralized. *Tier 4* is the actual kinetic studies done on a multicomponent media.

Anaerobic Tiered Testing

The philosophy of anaerobic tiered testing is the same as the aerobic. *Tier 1* is the use of a single carbon and energy source on an unacclimated, mixed anaerobic culture with the production of carbon dioxide and methane. Since it is possible to predict more or less the amount of gas produced if a substrate is completely mineralized, then mineralization can be deduced easily. Due to the slow growth of anaerobic bacteria, this stage can take a long time. *Tier 2* is the acclimation and enrichment stage. The test compound is increased in a multicomponent setting. Complete mineralization needs several interacting populations and furthermore, usually requires changes in the composition of the populations.

Testing for Recalcitrance

It is important that we can differentiate whether biodegradation has occurred or not. Intrinsically biodegradable compounds may be removed by abiotic mechanisms (e.g., sorption onto the biological floc or stripping due to aeration).[20] These are not desirable; stripping may be venting the toxin to another media (i.e., the air) and sorption onto the floc may result in a sludge that is unacceptable for disposal. Bearing these abiotic processes in mind, we should not necessarily believe long lists reporting biodegradability for various chemicals. Ghisalba[21] noted that the EPA's report "Water Related Environmental Fate of 129 Priority Pollutants" (1979) contains many contradictory statements and is suspect. Since this may have resulted from sorption and stripping and not biodegradation, it is important for us to be able to accurately measure biodegradation and furthermore, to be able to design the engineering system so that biodegradation can be maximized. Pitter[22] did experiments on many chemicals in order to assess their biodegradability. He used the substrate as a sole source of carbon for bacteria in activated sludge and calculated the percent chemical oxygen demand (COD) removal based on the initial concentration. He lists many chemicals and notes that biodegradation can be partial, acceptable, or total. This is an early study. While it does not directly address problems in the design system, it does conclude that there are many parameters that effect the reaction outcome. The physico-chemical aspects (e.g., temperature, solubility, the degree of dispersion of the compound in the media, pH, dissolved oxygen as well as the molecule size, chain length, kind and number of substituents, and stereochemistry) are as important as the biological factors (e.g., the microbiological culture, adaptation, age of the culture, toxicity of the compound, or other substrates). In order to reach a persistent, repeatable degradation of toxics, we must understand how these factors interact. The generation of intermediates, or end products and their effect on the treatment system has been an area of research. Chudoba reported[23] the adverse effects of the accumulation of waste product on bacterial activity. First there was a severe deterioration of the flocculating and settling properties of the mixed culture. Also there was a decrease in the COD removal. The waste product, proportional to the amount of substrate degraded, consisted of high molecular saccharidic polymers. There are still many questions that need to be answered. Can changing reactor configurations lead to a more stable culture? How do the dynamics of the different populations in a culture effect the system? What is the role of intermediates/end products? How can we best acclimate a culture? What about the toxicity of substances to be biodegraded?

Harman et al.[24] expressed dissatisfaction with past screening techniques that did not always include toxicity testing as a parameter. Such investigations failed to establish a primary cause of persistence. Since a large number of xenobiotics are persistent, this may be due to a lack of necessary enzymes by the microorganisms or inhibition by the chemical. Knowing the cause of persistence is important; if it is because of a lack of the appropriate enzymes we can choose either acclimation or bioaugmentation with microorganisms that do possess the enzymes. The authors were interested in what tests we can use to assess persistence and what operating conditions are favorable for the biotransformation of specific compounds. They outline testing procedures for orthochlorophenol (OCP) (highly inhibitory and moderately persistent) and 2,4-dichlorophenoxyacetic acid (DCP) (moderately inhibitory and slightly persistent) using bacteria from activated sludge for SBR's fed with potato starch wastewater. Biodegradation, toxicity, and metabolic activity tests were undertaken. The biodegradation tests were done by agar plating and shaker flasks. There was more growth of DCP than OCP. Acute toxicity tests were done with Daphnia and the Microtox tests, where OCP was more inhibitory. Metabolic activity tests were carried out by adding the compounds to bacteria and measuring the decreased oxygen uptake compared to a control. OCP showed the greatest decrease in oxygen uptake. Also cellular ATP levels were measured with OCP being found to have smaller levels. The study found discrepancies between individual tests and concluded that no one technique is useful for establishing persistence. To narrow the cause, acute toxicity tests were done and found OCP to be more toxic and inhibitory. To find the rate and degree of biodegradation a simulation test is needed (e.g., the batch shaker test). Use of the three tests will determine if a xenobiotic will persist in the reactor, the cause of persistence, and the estimated rate and degree of biodegradation that can be achieved. The metabolic testing is useful as a complimentary analysis.

Another point in question is if the technique we use to assess biodegradability really show this. In some instances where a high concentration of a chemical is required in order to achieve analytical precision, a toxic but biodegradable substance may be falsely labelled non-biodegradable due to its toxicity to the test organisms. Blok et al.[25] tested the sensitivity of different tests to a variety of chemicals. They suggest a method to minimize this false labelling of chemicals and provided insight into the testing procedures. Various chemicals were selected that were known to be biodegradable, had given variable results, but could be toxic. These chemicals were then tested for degradability via different techniques.

The **BOD_5 and closed bottle inhibition tests** establish toxicity by measuring the inhibitory effect of chemicals on the oxygen uptake resulting from the degradation of a readily degradable substrate (glucose/glutamic acid or a fatty acid ethoxylate). If a chemical is non-toxic and biodegradable, a higher oxygen uptake than the controls is expected.

The **activated sludge - inhibition of respiration test** measures the inhibitory effect of the test chemical on oxygen uptake of a respiring sludge while it is degrading a standard substrate at high concentration, thus giving a high respiration rate.

The **growth inhibition test** is a measure of a decrease in the turbidity of microorganisms in solution as their growth is stopped because of the addition of a toxin. The Microtox measures a light reduction by the luminescent bacterial species *Phytobacterium phophoreum* due to a toxic presence. In the **repetitive die away test** two cultures are incubated, one with a growth substrate alone and one with the test chemical. Toxicity is a measure of the reduced oxygen uptake.

The tests were compared by an EC50, that is the concentration of the substrate that causes a 50% reduction in the specific test parameter. Microtox is the most sensitive of the tests and may overestimate the toxicity of a chemical. The inhibition of respiration test proved to be consistent and may be good for testing where there are high innoculum concentrations. The growth inhibition, RDA, and BOD tests seemed to have similar sensitivities. From the chemicals tested, no single test gave EC50 values that were consistently related to toxicity. It was found that in most cases, the EC50 concentration were inhibitory. Therefore, the authors suggest that the test substance concentrations used in biodegradability testing should be less than one-tenth that of the EC50 values obtained in toxicity testing.

PILOT STUDIES

PCB Biodegradation

Polychlorinate biphenols (PCB) are very toxic chemicals and suspected carcinogens. They include polychlorinated dibenzodioxins (PCDD), polychlorinated dibenzofurans (PCDF), and chlorinated benzenes. They have been found in sediments where they are a threat to the drinking water. A consortium of bacteria has been isolated that appears able to degrade PCBs. Some strains of bacteria were isolated that had varying degrees of competence in degrading PCBs. The bacteria metabolized the PCBs by using enzymes specific to the strain. In many cases a particular isomer may not be completely degraded, and a second species is needed to catalyze the degradation of the intermediate (although it cannot degrade the parent compound). Some 25 strains of indigenous, and chemically mutated (grown naturally in diverse PCB sediments) have been isolated and tested. Of these *Alcaligenes eutrophs* and *Pseudomonal putida* degraded 13 different isomers to the extent of between 80 and 100%. Also *Arthrobacteria* were found to degrade most isomers. The degradation pathway may occur via oxidation of PCBs to chlorobenzoic acid. *Alcaligenes* are believed to have a dioxygenase that attacks the carbon at position 3,4. *Corynebacterium* sp MBl is believed to have a 2,3 dioxygenase. Effective strains usually contain plasmids. A mutant strain that has lost its plasmid can no longer metabolize PCBs. A few strains mineralize some chlorinated biphenols but in most cases a strain can degrade one ring of the PCB, but are unable to degrade the resulting chlorobenzoates which accumulate. *Pseudomonas* sp B13 is a hybrid that has the relatively non-specific oxygenase of *P.putida* and can completely mineralize the chlorobenzoates. In general, the more heavily chlorinated structures resist degradation. The mechanism of hydroxylation of PCBs has not been elucidated, nor an enzyme isolated. However, two pathways have been proposed.

The first involves initial hydroxylation in the 2,3 position of the less substituted ring, followed by meta cleavage and subsequent degradation of the alphatic portion to form unsubstituted benzoic acids. The chlorines on the aliphatic carbon are lost in this process. This may not be the mechanism for the degradation of PCB's substitutes in all the orthopositions. A second pathway may be the preferential 3,4 dioxygenase of *A.eutrophus* H850 or the 2,3 dioxygenase of *Corynebacterium* sp MBl.

The U.S. EPA has written about a biological process developed by Bioclean Inc. called the Bioclean Naturally Adapted Microbial Process.[26] Here the microbes can be bacteria or fungi; the method being aerobic, anaerobic, or facultative. This has been successful with cleaning pentachlorophenol bearing sediments. The process was evaluated along with abiotic mechanisms for the decontamination of PCB bearing sediments and was found to be quite useful. The bioclean process does not need pretreatment of the sediments and does not appear to generate any RCRA wastes or emissions. It involves two steps:

1. The extraction, sterilization, and solubilization of the contaminants using high pH and temperature.

2. The bacterial destruction of the contaminant.

These take place over a three day period and have the capacity for treating 22-28 m^3/day. The processing cycle has the following stages:

1. The contaminated sediment is fed to a digester in a slurry if that is how it is received. The final digester charge is 2/3 water and 1/3 solids. It is made alkaline by the addition of sodium hydroxide and then the temperature raised to 82°C for one hour. The mix is agitated throughout.

2. It is extracted, cooled to 30°C, and neutralized. Bacteria is inoculated and left for 48-72 hours. The process is aerobic so sterile filtered air is fed through.

3. The treated batch is discharged to a dewatering pit where the sediment is separated for redeposition.

4. A decontamination period follows after which Stage 1 can begin again.

Under SITE, the "superfund innovative technology evaluation" scheme developed to accelerate new hazardous waste treatment technologies, Detox Industries (Texas) reported a way to treat PCE, creosote, oil, phenolics, and pentachlorophenol in-situ.[27] They culture mutant microorganisms that feed on the target contaminant and then apply them to the site in an aqueous solution. The compounds are slowly metabolized over a period of several months. Ultimately the bacteria die from lack of nutrients. Problems with the process are that heavy metals can inhibit the growth of the bacteria, but Detox states that 14,000 cu. yd. of PCB contaminated soil was reduced from levels of 2900 ppm to 1 ppm at 20% of the cost of excavation.

Kampbell et al.[28] investigated the use of a soil bioreactor for the degradation of isobutane, butane, and propane and found that the mineralization was rapid and extensive. At higher concentrations the biodegradation rate was limited by the microbial capacity to metabolize the organic compound. At 90 cm depth they found less biodegradation and they linked this to the liner which was used to protect the groundwater from being contaminated, but kept the soil wetter. So better drainage is needed. The probable intermediates are alcohols and ketones which are produced through the hydroxylation of the alkenes by monoxygenase bacterial enzymes which start alkene metabolism. These are readily degraded in oxygenated soil. The hydrocarbons were completely degraded without any accumulation of the propellant hydrocarbons or intermediates in the leachate. The overall efficiency of the bioreactor is 95-99% at 24°C although the acclimation time was slow. The ability to degrade volatile organic compounds decreased at low temperatures. But for soil temperatures between 12°C and 24°C, the bioreactor functioned well. The study showed an average transformation of greater than 90% for the volatile organic compounds between 1984-1985. If the temperature is sustained, then there could be good biodegradation all year. The bioreactor removed TCE as rapidly as the most refractory propellant hydrocarbon. The study showed that organic compounds can be removed from the air by permeation through soil bioreactors. Further work is needed to determine the most appropriate texture and type of soil, to see if the addition of mineral nutrients will enhance biodegradation, and to assess the effect of flow rate, bed depth, organic vapor concentration, and soil moisture content on the system. The predicted behavior of the bioreactor based on laboratory studies agreed closely with the actual behavior of the field system.

Methyl Ethyl Ketone

Skladany et al.[29] found in-situ treatment of methyl ethyl ketone (MEK) to be successful. Detox industries supplied two submerged bioreactors to a plant that had MEK as a waste product. The first was to act as the biodegradation unit and the second as the polishing unit. Bacteria were to be grown on the fixed plastic film and supplied oxygen, organic carbon, the inorganic nutrients nitrogen and phosphorous, and a pH range of 7-8. The reactors were put in place and given four weeks pretreatment before MEK was fed in batch mode. They found that the MEK was degraded to such an extent that the bacteria would no longer have a growth substance and so additional waste producers were fed to the reactor. At this point a heat exchanger was installed in the reactor in readiness. The compounds were cyclohexanone, miscellaneous organics, and miscellaneous solvents. The cyclohexanone was readily biodegraded, the MEK continued to be removed to >99% and the others degraded slightly less. The system was sensitive to shock loadings and care had to be taken in loading the compounds. However the in-situ overall biodegradation of MEK and other compounds proved to be successful. The reactors were in operation between May 1986 to May 1987 and 3,092,919 gallons of wastewater were treated.

A patent[30] was granted to Research Co., San Francisco, CA for an in-situ process for the biodegradation of hydrocarbon soil where oxygen is pumped into the soil using a borehole in the earth's surface having hydrocarbon-degrading microbes in place.

Landfill Leachate

Ying et al.[31] investigated the feasibility of adding powdered activated carbon (PAC) to sequencing batch bioreactors already in place for the treatment of landfill leachate from the Hyde Park landfill, Niagara, NY instead of a subsequent PAC treatment. Since many organic compounds are readily absorbed on to PAC, the authors proposed that biodegradation would be enhanced because of the long residence time of the compounds on PAC. This was a successful pilot scale study where improvements were found in the effluent quality, with more organic removal and better sludge settling with better dewaterability, and the bacteria being more resistant to shock loading of the wastewater. The effluent needed treatment with only 4% of the carbon which would have been needed for treating the raw leachate alone. Nitrification and denitrification were observed in all bioreactors with no aeration during the last two hours of treatment so the PAC-SBR treatment can accomplish nitrogen removals as well. So a high quality effluent was produced compared to the two stage biodegradation followed by activated carbon treatment. Additional work by Weber et al.[32] studied the effect that an intermittent treatment of a SBR-PAC reactor would have on degradation rates. A pilot study was initiated to test the feasibility of the process. A former hazardous waste site was found to be contaminated by a wide variety of hazardous organic substances. Since there was a projected low groundwater pumping rate at this site, the on-site treatment was expected to be most cost effective if carried out on an intermittent basis. They found that the rate of COD removal was adversely affected by decreases in operation frequency and decreasing operation temperatures. But all the biological BOD was removed within a reaction period of 24 hours despite periods as long as three weeks between operations. The addition of PAC was advantageous for decreasing the aeration times, for conditions of low temperature and extended lag periods. This pilot study showed that intermittent biological treatment can achieve significant reduction of waste pollutants. COD removal was decreased by decreasing operation frequency but if the reaction periods are extended, the biodegradable

COD can be removed in bioreactors having long lag periods. The intermittent process has potential applications for leachate treatment of municipal and hazardous waste landfills, industrial waste flows released on an intermittent basis, and other waste sites needing remediation.

LABORATORY STUDIES OF AEROBIC DEGRADATION

There has been much research on aerobic degradation in the laboratory. These are at both the experimental and bench scale level. While there is a long way to go before these procedures are ready to be used at treatment plants, or in-situ treatment, they are a necessary part of the learning process that may eventually lead to a feasible biological treatment method for a particular hazardous waste. Included in this section are reports about the biodegradation of TCE, chlorophenol, TCDD, PNP, phenanthrene, and chlorinated groundwaters.

TCE Degradation

Trichloroethene (TCE) is a common contaminant of our environment. It has been used extensively in the past in the dry cleaning industry for example. It is generally resistant to aerobic degradation. An anaerobic process will be described later. This anaerobic process has as its end product, vinyl chloride, a more potent carcinogen than TCE, and hence a different degradation pathway would be better. Some methanotrophic bacteria which utilize methane have been found to degrade TCE. This is important since it is an aerobic process and one that might be able to be used in-situ. Little et al.[33] isolated some strains of bacteria from a waste disposal site near Oak Ridge, TN. It had been used for direct dumping of chlorinated organic solvent wastes for a number of years. The cultures were grown and the pure methanotrophic strains isolated. These were called 46-1 and 68-1. These appeared to be Type 1 methanotrophic bacteria based on their enzymatic activities and internal structure. Type 46-1 was chosen for the experiments due to its growth characteristics. In pure culture Type 46-1 converted 40% of TCE in a methane limited environment, incubated for twenty days. Most of the converted TCE was water soluble breakdown products, although some fraction was completely mineralized to carbon dioxide. It was found that the cells stopped growing and degrading TCE once the methane had been used. When the experiment was repeated but with the addition of more methane, more TCE was degraded. The fact that methane or methanol is needed by the bacteria suggest a cometabolic process. However, it seems that the methanotrophic bacteria alone cannot completely degrade TCE, but that the heterotrophic bacteria normally associated with it finish the process. The apparent accumulation of glyoxylic acid, dichloroacetic acid, and carbon dioxide in TCE degrading cultures suggests a possible mechanism of biodegradation. The authors propose that the methane-oxidizing bacteria, such as Strain 46-1, convert TCE to its epoxide, which then breaks down spontaneously in water to form dichloroacetic acid, glyoxylic acid, or one-carbon compounds such as formate or carbon monoxide. The two carbon acids accumulate in the water phase, while formate and carbon monoxide are further oxidized by methanotrophic bacteria to carbon dioxide. The hydrolysis of the TCE epoxide to form one carbon products explains the production of carbon dioxide from TCE by this methanotrophic organism. The degradation pathway can be seen in the figure that follows.

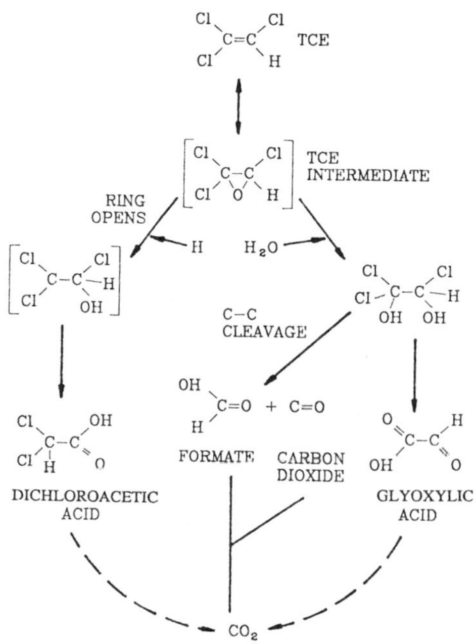

Figure 3. Reprinted from: TCE Degradation by a Methane-Oxidizing Bacterium. Little et al., App & Env Mic, April 1988.

The authors suggest a potential for treating sites contaminated by chlorinated alkenes with methanotrophic bacteria.

Polycyclic Aromatic Hydrocarbon Degradation

There have been numerous studies concerned with PAHs. These occur as natural constituents and combustion products of fossil fuels and are widespread environmental contaminants. They are toxic and have a potential for bioaccumulation and many are carcinogens. The most important mode of degradation may be via microorganisms. The ability to degrade PAHs depends on the complexity of their structure and the extent of enzymatic adaptation by bacteria. In general PAHs with 2 or 3 aromatic rings are readily degraded, but ones with 4 or more are usually recalcitrant and genotoxic.

Heitkemp et al.[34] isolated *Mycobacterium* sp, which readily degraded pyrene in pure culture under optimal conditions. This species was found in an ecosystem chronically exposed to petrogenic hydrocarbons. They treated both induced and non-induced cultures with chloramphenicol and showed that pyrene degrading enzymes were inducible in this *Mycobacterium* sp. This species could also degrade other PAHs and alkyl and nitro substituted PAHs such as naphthalene, phenanthrene, fluoranthenem 3-methylchloranethene, 1-nitropyrene, and 6-nitrochrysene.

This is the first report of a bacteria being able to extensively mineralize pyrene and other PAHs with four aromatic rings. Further investigation,[35] using radioactive carbon tracers elucidated the pathway for the microbial catabolism of pyrene. Over 60% of pyrene was mineralized to carbon dioxide after 96 hours when incubated at 24°C. The metabolites were cis- and trans-4,5 pyrene dihydrodiols and pyrenol (as the initial microbial ring oxidation products of pyrene). The major metabolite was 4-phenanthroic acid and 4-hydroxyperinapthenone. Cinnamic and phthalic acids were identified as ring fission products. Studies showed that the formation of the cis- and trans-4,5 pyrene dihydrodiols were catalyzed by dioxygenase and monoxygenase enzymes respectively. The initial oxidative attack on pyrene resulted in the formation of dihydrodiols at the 4,5 position. The inducible enzymes responsible for the degradation could be chromosomal or plasmid in origin. If we can isolate the genetic material that codes for the enzymes, future degradation in-situ may be possible.

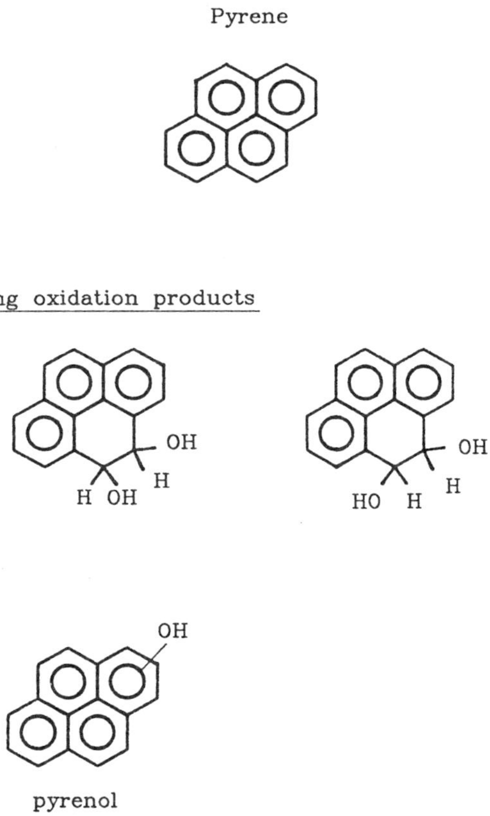

Figure 4. Ring oxidation products.

Ring Fission Products

PAHs found that the interactions were important to the rate and total extent of degradation of individual PAHs. Exposure of the marine sediments to a particular PAH or benzene enhanced the ability of those sediments to degrade that PAH as well as others.

They enriched bacteria with individual PAHs or combinations of them and then followed this with incubation with PAHs. They found that exposure to 3 PAHs and benzene led to an increased amount of subsequent anthracene. Prior exposure to phenanthrene led to a rise in

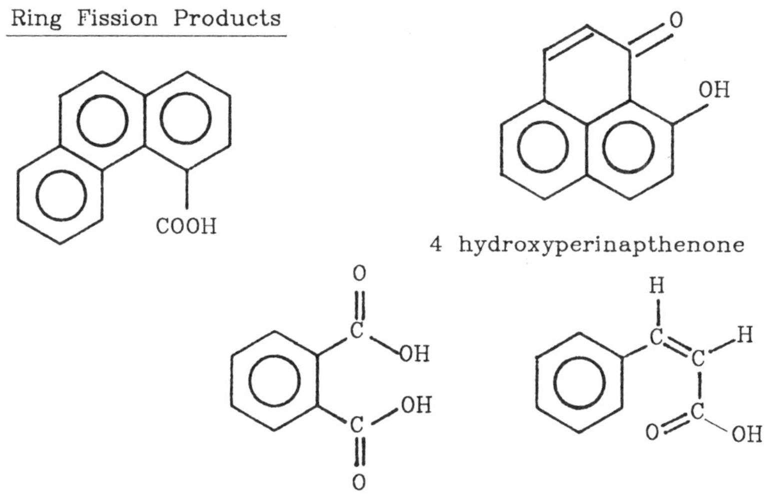

Figure 5. Ring fission products.

the initial degradation of anthracene. Prior exposure to napthalene led to a rise in the initial degradation of phenanthrene but not anthracene. Napthalene mineralization was enhanced by pre-exposure to all other aromatics after 14 days. It is suggested that there is either a less specific or more degradative mechanism for the compound. This phenomenon may be because of the selection and proliferation of specific bacterial populations. The enhanced degradation of other PAHs suggests that the population selected has either a broad specificity to the PAHs with perhaps functionally similar oxidases or common pathways of PAH degradation or both. There may also be a broad range of populations selected and within this, different bacteria each having some ability to degrade PAHs.

Interactions between PAHs may be another factor governing the rates of degradation of such hydrocarbons in sedimentary environments.

Mihelck et al.[37] looked at several PAHs under aerobic, anaerobic, and dentrifying conditions in soil-water systems. The compounds were napthalene, acenaphthene, and napthol which all were degraded by bacteria in aerobic conditions in 10, 10, and 3 days, respectively. They found that under anaerobic conditions napthol was at non-detectable levels after the fifteenth day but that napthalene and acenapthene showed no significant degradation over 5-70 days. In the dentrifying experiments, all were degraded to non-detectable levels in 45, 40, and 16 days, respectively. An acclimation period of two weeks was necessary. They also tested naphthol and naphthalene with the abiotic reaction with maganese oxide and found that napthol showed no degradation over 9 weeks, whereas naphthalene decreased to non-detectable levels in 9 days. Napthol was degraded under all conditions. The authors conclude that this is because of the hydroxyl group which makes it more reactive. An acclimation period of two days for aerobic degradation and a period of two weeks for denitrifying conditions were observed for both naphthalene and acenaphthene. This is the first work showing that low weight unsubstituted PAH are amenable to degradation under dentrifying conditions. Therefore microbial restoration of anoxic soil sediments and groundwater systems contaminated with PAH compounds is possible.

It is possible that other factors play a role in microbiological degradation. Mahaffey et al.[38] studied the degradation of benz(a)anthracene. The *Beirjerinkia* strain B1 used in this study was unable to utilize benz(a)anthracene as a source of carbon and energy for growth.

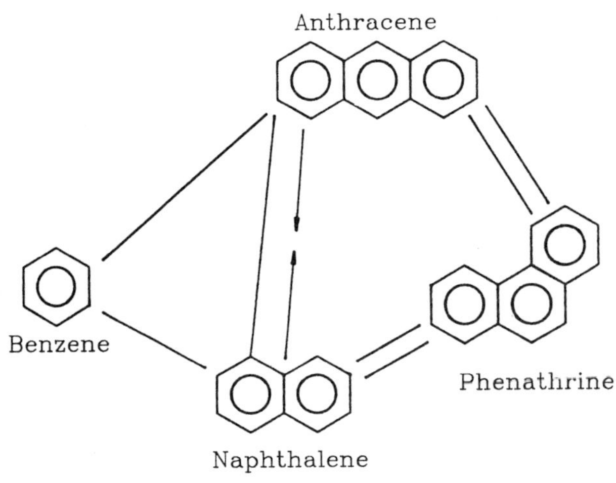

Figure 6. Proposed relationship between benzene and the PAHs used in the study.

However after growth with succinate in the presence of biphenyl, m-xylene, or salicylate as inducing substrates, the strain oxidized benz(a)anthracene to carbon dioxide and a mixture of o-hydroxy polyaromatic acids. The cells induced with biphenyl were the most active degraders. 1-hydroxy 2-anthranoic acid was the major product. The two minor metabolites were 2-hydroxy 3-phenathroic acid and 3-hydroxy 3-phenanthroic acids.

Beirjerinkia strain B8/36 converts most of benz(a)anthracene to the cis-1,2 diol (see Figure 7).

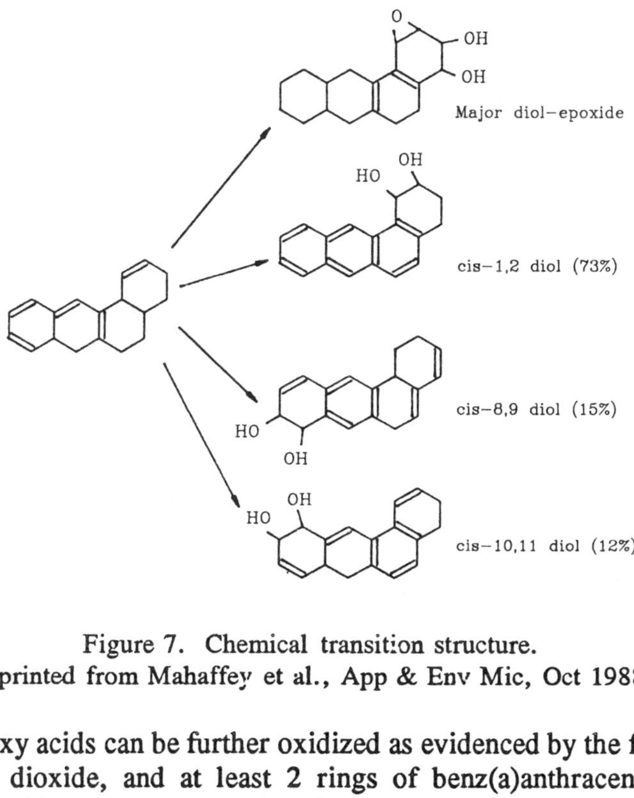

Figure 7. Chemical transition structure.
Reprinted from Mahaffey et al., App & Env Mic, Oct 1988.

The hydroxy acids can be further oxidized as evidenced by the formation of carbon dioxide, and at least 2 rings of benz(a)anthracene can be degraded. Since these molecules do not appear to induce their own metabolism, cooxidation may represent a significant mode of degradation in soil.

Bauer et al.[39] also found that the polychaete *Capitella capitata* has degrative abilities for PAHs, e.g., anthracene mineralized after 5 months.

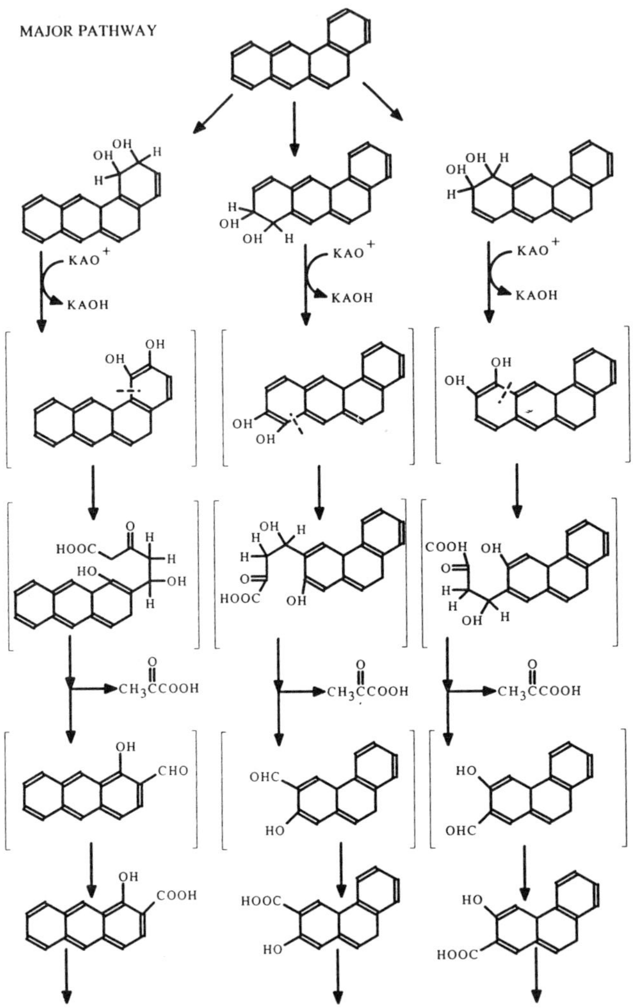

Figure 8. Proposed pathways for the metabolism of benz(a)anthracene by Beijerinckia strain B1. Reprinted from Mahaffey, et al., App & Env Mic, Oct. '88, p. 2421.

Phenanthrene Degradation

Polycycloaromatic hydrocarbons in the environment are a great concern because of their toxicity, carcinogenicity, resistance to biodegradation, and because they are ubiquitous. Phenanthrene was chosen by Guerin and Jones[40] because although it is not particularly toxic it is a good model compound for studying PAH degradation. They studied cultured bacteria with two concentrations of phenanthrene, with quite different results. In both cases, the intermediate product of 1-hydroxy 2-napthoic acid (1H2NA) was obtained.

At low concentrations, 1H2NA accumulates to a lesser extent than at high concentrations. It is thought that when phenanthrene is present in excess, it acts as a readily available source of carbon and the cultures are dominated by the phenanthrene degraders. The phenanthrene is degraded to 1H2NA. The 1H2NA is produced more rapidly than it is consumed in this first stage and so it accumulates. Only when the phenanthrene is depleted does the 1H2NA degrade, allowing a second stage of protein production to start. At low concentrations, the phenanthrene degraders mineralize the substance more efficiently so that the intermediate doesn't accumulate. This two stage degradation pattern may also be the method used for the degradation of naphthalene, anthracene, benzanthracene, and BAP. The authors provide a note of caution. The transformation of PAH to polar intermediates has implications in measuring the rate of degradation, since the disappearance of the parent compound does not mean that it is completely mineralized.

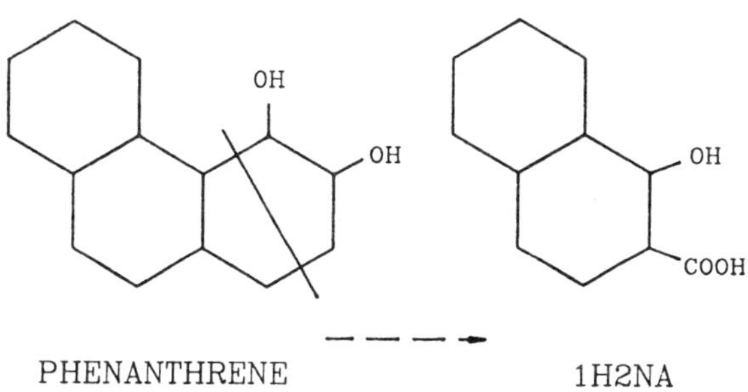

Figure 9. Phenanthrene degradation.

Chlorophenol Degradation

An interesting biodegradative technique, involving the aquatic macrophyte pennywort (*Hydrocotyle umbrellata*) and bacteria was reported by Dieberg et al.[41] In this study, they placed a thin layer of pennywort in a raceway and fed through sewage with 2,4-dichlorophenol

(2,4-DCP) and monochloramine added to it. They found an acclimation to chlorophenol which suggests that the subsequent degradation was biotic. The plants developed masses of roots which together with the colonizing bacteria removed the toxins to extremely low concentrations; 100% for monochloramine and >95% for 2,4-DCP, even with high loading rates and low residence times over the sixty day study. Pennywort only needs harvesting or thinning one or two times a year. Removal of contaminants such as 2,4-DCP and chloramines, can therefore be attained in pennywort thin film systems without incurring a major waste problem. The authors believe the phenol was primarily degraded to other compounds by the plant. They did not isolate the bacteria present, nor assess if the bacteria are necessary for the biodegradation; however, it is one of the rare studies involving plants as the degradative agents.

Chlorinated Wastes

Highly chlorinated wastes are generally not considered readily biodegradable. Contaminated water may contain many of these carcinogens. Pelon and Mayo[42] wanted to see if bacteria would utilize various chlorinated wastes as the only carbon source. The wastes included, hexachlorobutadiene, 1,1,2-tricloroethane, tetra-chloroethane, hexachlorobenzene, and 1,2-dichloroethane. Bacteria from soils at the dump site were cultured with these organics with the only addition of salts. The experiments used aerobic mixing and a temperature of 25°C. They reported that the bacterial cultures were able to grow on the organic saturated groundwater with no added carbon source. They noted that the populations changed both qualitatively and quantitatively. The authors did not determine if indeed any of the compounds were mineralized, or if intermediate products were formed and they only hint at the presence of a consortium. Further information is needed, but the study does show that some bacteria can utilize these wastes.

p-Nitrophenol Degradation

Many factors limit the ability of bacteria to degrade compounds. If we are to treat hazardous wastes in a reliable manner, we need to know more about these factors. Yaldi et al.[43] were interested in factors that limit the ability of bacteria added to lake water to stimulate biodegradation. *Corynebacterium* sp were used for the study. They were made resistant to antibiotics by growing them in increasing concentrations of streptomycin, kasugamycin, and spectinomycin. The test compound was p-nitrophenol (PNP) which was added to water collected from lakes. The rate of biodegradation for low initial concentrations of PNP was found to be less than expected when extrapolated back from higher concentrations of PNP. The authors found that this was in fact due to inorganic deficiencies in the lake water collected at certain times. Therefore, the degradation of even a low concentration of PNP was not good. When there had been no rainfall, nutrients were not leached from the soil into the lake water. By adding nitrogen and phosphorous to the water, mineralization was increased. Also the addition of glucose as a second substrate, did not reduce or delay mineralization but actually increased it to some 70%. When more phosphorous was added, the mineralization was even higher at 82%. This study shows how other factors interact with the bacteria, both the nutrients needed and the population dynamics play a role.

Degradation of Fluoro Substituted Benzenes

Renganathen[44] found that *Pseudomonas* sp strain T-12 cells in which the toluene degrading enzymes have been induced can transform many 3-fluoro-substituted benzenes to the corresponding 2,3-catechols with the concomitant release of inorganic fluoride. The substrates that induce 2,3-dioxygenase are 3-fluorotoluene, 3-fluorotrifluorotoluene, 3-fluoro-halobenzene, 3-fluoronisole, and 3-fluorobenzonitrile. While 3-fluoro-toluene and 3-fluoronisole produced only deflorinated catechols, other substrates led to catechol products both with and without the toluene substituent. The stearic size of the C-1 substituent plays an important role in determining the deflourination reaction. The probable mechanism for defluorinating involves toluene 2,3-dioxygenase (see Figure 10).

Pentachlorophenol Degradation

Topp et al.[45] investigated the interaction of pentachlorophenol (PCP) and an alternate carbon source on the survival and activities of *Flavobacterium* in axenic cultures. They found that PCB degradation does not require extra carbon source but the acclimation time is decreased if it is present. Cultures raised on PCB alone have activities

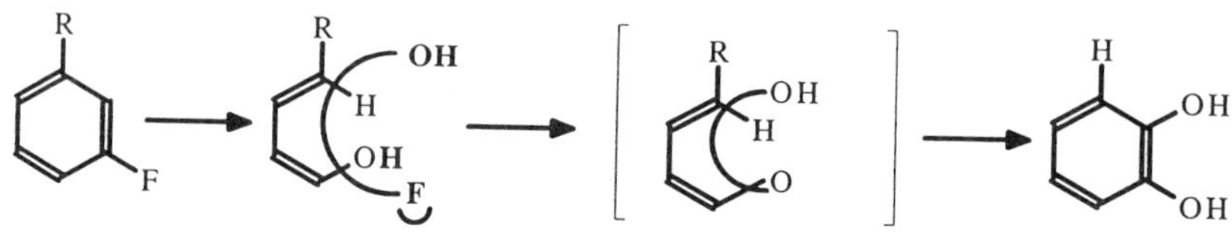

Figure 10. Possible mechanism of defluorination of 3-fluoro substituted benzenes by *Pseudomonas* sp, strain T-12 involving toluene-2,3-dioxygenase. Reprinted from Renganathan, App & Env Mic, vol. 55 #2, Feb. '89, p. 332.

slightly greater than with supplementary carbon. The decrease in lag time before degradation occurs could be due to a reduction in induction time or the stimulation of an activity of an already induced population or both. The cells were sensitive to shock loads of PCP but the survivors had more resistance. When PCP induced cells grown on PCP and sodium glutamate were removed to PCP substrate only, there was a decrease in biodegradation. Adding sodium glutamate fully restored the activity. This study has important implications for in-situ treatment, where cells grown in-vitro and then exposed to high concentrations of PCP in-situ may survive and degrade the PCP better if there is metabolizable carbon in the environment as a supplementary source of food.

Oil Degradation

Another major source of waste is from the oil industry. Using bacteria to degrade oil is in the experimental stage, but further research[46] may prove its utility in removing oil from the environment. Feld et al.[47] describes the biotransformation of oil based drill cuttings and a production sludge from an oil field waste pit. These are not usually biodegradable. An emulsification and demulsification process are critical and convert the non-biodegradable compounds to ones that are biodegradable. They measured the respiration rates in a batch reactor and found two microbial cultures from the environment that would degrade the batch drill cuttings and the sludge. They found that the concentration of the sludges was important; at too high a concentration the mixture could not be mixed and properly aerated so that the bacteria could not utilize the source. To achieve dewatering of the sludge, polymer additions are required, but after using bacteria, less and less polymers had to be added to achieve the same effect and hence bacteria may be useful as a pretreatment for the dewatering stage. A pilot test is planned. Emulsan is the extra cellular form of a heteropoly-saccharide produced by the oil degrading bacterium *Acinebacter calcoaceticus* RAG-1. It has potential use in the oil industry for example in reducing oil viscosity during pipe line transportation. Foght et al.[48] studied the effect of emulsan on the biodegradation of oil by both pure cultures of bacteria and mixed cultures. They found a 50-90% decrease in degradation of alkanes after treatment with emulsan. The aromatic compound degradation was largely unaffected. The inhibition may be due to the emulsan stopping a direct physical interaction between the bacterial cells with the hydrophobic media that is necessary to initiate biodegradation. Several mechanisms are proposed; it could be due to a masking of hydrophobic sites on the bacterial cells. Further research is required if we are to successfully overcome this obstacle to biodegradation after oil is treated with emulsan. Although the emulsan makes transportation of oil easier, at present the choice of remediation by bacterial degradation in the case of a spill is weakened. Use of an emulsifying agent produced by a *Pseudomonas* strain has been demonstrated[49] in field trials in the Middle East where when added to oil storage tanks for four days, it released 5600 barrels of oil that would normally have been lost and reduced the waste.

Hexachlorocyclohexane Degradation

Gamma-hexachlorocyclohexane (HCH) or lindane is a widely used pesticide. During its production, a mixture of isomers are formed. The most important of them are alpha, beta, gamma, and delta. The gamma form is the only effective insecticide. Bachman et al.[50] tested HCH breakdown from contaminated soils in the Netherlands aerobically, in methanogenic, denitrifying, and sulfate reducing conditions. This study found that HCH was better degraded in aerobic conditions, contrary to results in earlier literature. In aerobic conditions, HCH was mineralized in 18 days with the only intermediate detected being pentachloro-cyclohexane. The methanogenic bacteria mineralized HCH at a slower rate, with 85% of the breakdown products identified as monochlorobenzene, 3,5-dichlorophenol, and a trichlorophenol, possibly 2,4,5-trichlorophenol. In sulfate reducing and denitrifying conditions, no significant degradation of HCH was found to be superior, and care must be taken to prevent anaerobiosis in HCH contaminated soils. A later study by De Bruin et al.[51] studied the effects of temperature, auxiliary carbon, oxygen, substrate concentration, and availability on the degradation of alpha HCH in aerobic conditions. We need an understanding of these factors, for the development of efficient strategies for the clean-up of contaminated waste. They found that temperatures in the range of 20-30°C were the most favorable. The addition of an auxiliary carbon source repressed the HCH degradation, possibly because the bacteria preferentially used the more easily degradable carbon source. Increasing the oxygen partial pressure reduced the repressive effect. They found there was a linear relationship between the concentration of alpha HCH and its conversion rate up to concentrations of 900 mg HCH/kg, after which there was a saturation effect. The alpha degrading enzymes were apparently already induced in the contaminated soil slurry. Inhomogeneities were found to have an important role in HCH degradation. The soil had been sieved and homogenized prior to use in the experiments, but when more alpha HCH was added to try and overcome the lag effect, they found that although the lag phase disappeared, the biodegradation rates decreased. The authors hypothesized that this was because the not very soluble alpha was creating more inhomogeneity. When they added the more soluble gamma form, the lag phase disappeared and the rates of biodegradation increased. So, the solubilization and adsorption-desorption characteristics of HCH were important in the biodegradation of this chemical. This observation is important for the application of biotransformation methods for cleaning up contaminated soils and interpretation of laboratory studies quantifying biodegradation. The conversation pathway is believed to be as shown in Figure 11.

The study does not show if one or more pathways occur concomitantly. Bergman et al.[52] studied the degradation of alpha and beta HCH under anaerobic field conditions and found that even the relatively easy to degrade gamma HCH was not degraded. They assess the failure as being due to the environmental conditions of the soil plus the low temperatures. This suggestion is supported by the work described above on HCH.

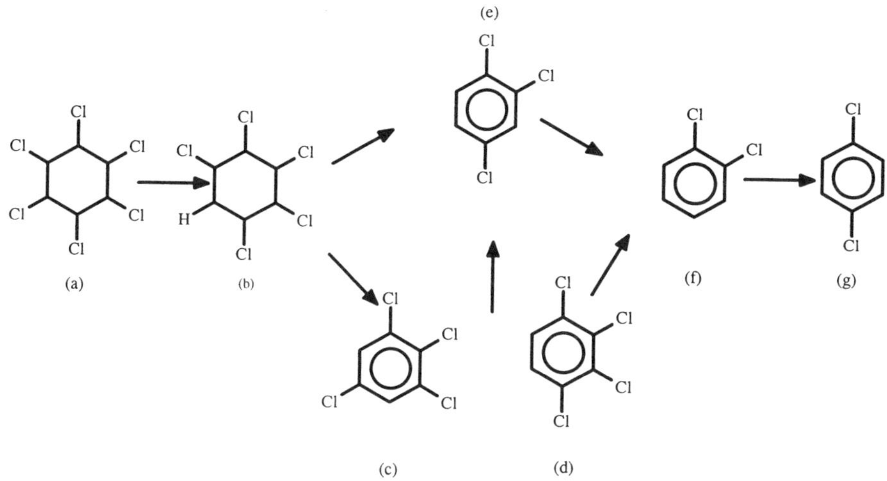

Figure 11. Proposed initial steps of alph-HCH degradation under aerobic conditions in a slurry of contaminated soil. (A) Alpha-HCH; (B) PCHH; (C) 1,2,3,5-TeCB; (D) 1,2,4-TCB; (E) 1,2,4-DCB; (F) 1,2-DCB; (G) 1,4-DCB. Reprinted from Bachman et al., App & Env Mic, vol. 54, Feb. '88, p. 553.

Aniline Degradation

In the environment not only one carbon source is usually available. This alternate substrate may either promote or inhibit the utilization of the chemical of concern. Konopka et al.[53] was interested in the effect that a secondary carbon had on the degradation of aniline by a *Pseudomonas* sp that was isolated from sewage. Anilines are used in the manufacture of dyes, plastics, and pharmaceuticals. The strain K1 had a narrow range of aromatic substrate utilization. Resting cells from aniline grown cultures exhibited high respiratory activity upon the addition of aniline or catechol, some activity with toluidine, and no activity with the addition of a wide variety of other aromatic compounds, including dihydroxybenzylamine, chloroanilines, ethylanilines, aminophenols, aminobenzoates, and dihydroxybenzoates. The enzymes responsible for the degradation of aniline have some 2,3 dioxygenase activity. This was an inducible system, as was shown in the inability of cells grown on lactate and acetate to respire when given aniline and by the absence of catechol dioxygenase in those cells. The simultaneous utilization of aniline and other organic substrates by strain K1 has significance for the degradation of aniline in natural environments. If other substrates such as glycerol or lactate are present, this strain can continue to degrade aniline, and if the aniline is present only in small concentrations, the microbial population can still be supported which can then transform the xenobiotic at a high rate.

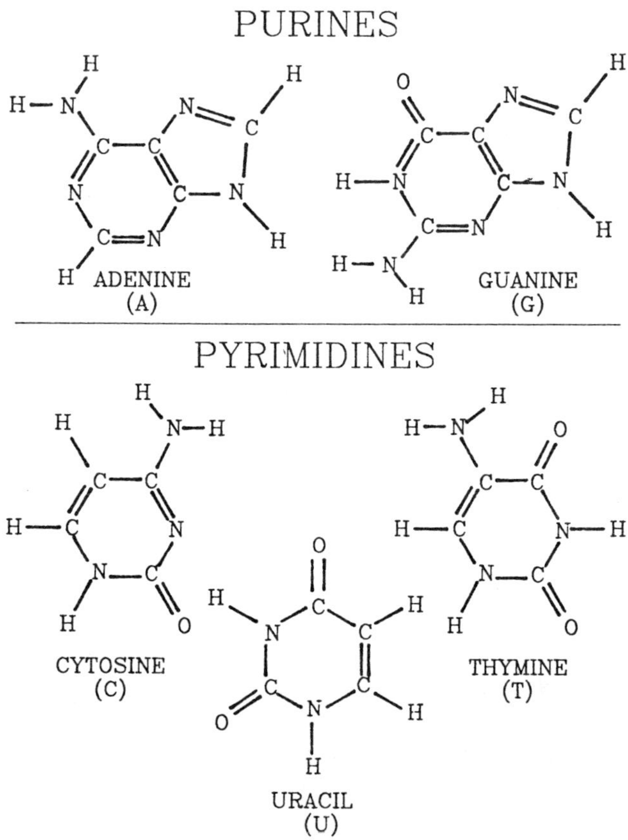

Figure 12. Purines; pyrimidines -- the five bases.

Disulfide Removal

Contamination does not always involve the groundwater or soil systems, although much of the work has focused on these media. Kanagwa and

Mikami's report[54] is how *Thiobacillus thioparus* successfully removed various disulfides from the air. Methanethiol (methylmercaptan) (MM), dimethyl sulfide (DMS), dimethyl disulfide (DMDS), and hydrogen sulfide are malodorous compounds produced by the wood-pulping industry, in oil refineries, and in manure and sewer systems. They were found to be removed by the bacterium sometimes far more rapidly than with acclimated sludge.

Metolachlor Degradation

A study by Zheng et al.[55] investigated the nature and extent of sorption and metabolism of the herbicide metolachlor by a bacterial community as opposed to one species. They found that a mixed bacterial culture (J4-A) exhibited a remarkable capacity to take up and accumulate metolachlor from a liquid medium and in addition possessed a metabolic potential for transforming the herbicide. The pure culture (B-2), although able to accumulate metolachlor, was much less effective than the mixed population. The synergistic action among members of the microbial community apparently led to better growth and consequently increased accumulation and metabolism of the herbicide. The metabolic pathway involves dechlorination, since the products found included dechlorinated metolachlor. The intentional seeding of bacterial populations may be useful in the removal of this herbicide and its metabolites from the environment.

Activated Sludge Studies

The activated sludge process is a widely used process for the removal of organic and inorganic substances from industrial and municipal wastewaters. A problem with this process is the presence of inhibitory or toxic substances in the wastewater. Heavy metals in particular can upset the process. Hexavalent chromium is a heavy metal that has to be removed from the sludge process. Lee et al.[56] wanted to assess the effect of adding powdered activated carbon (PAC) to the activated sludge process for treatment of wastewaters containing Cr(VI). The adsorption of Cr(VI) on activated sludge was not as high as for PAC which had a very high initial adsorption rate; 66% of the equilibrium value was reached in 30 minutes. There were high removal efficiencies of COD and Cr(VI), with 41% removal of chromium in the PAC unit compared to the control. The addition of PAC also led to a decrease in the time that the unit required to recover from stress. The addition of the PAC to the aeration unit increased overall microorganism growth rate and biological removal rate of the substrate.

Grosse et al.[57] report on the results of a bench scale study of activated sludge degradation of methyl ethyl ketone (MEK) and 1,1,1-trichlorothane (TCA). Each system consisted of a completely mixed, continuous flow reactor and two clarifiers. They were acclimated to a synthetic feed for one month before introduction of the sludge. There were several system upsets that had to be overcome; an operational failure stopped the test and shock loads of TCA had a detrimental effect on the stability of the sludge. Both the MEK and TCA systems experienced secondary sludge and recycle problems attributable to the character of the sludge formed and the design of the secondary clarifier. The sludge flocs formed were fine and light and did not settle well. Floc formation may be associated with the use of the synthetic wastewater. Nutrient deficiencies may also have been a factor in the operation of the MEK system at the higher concentrations. They found that while biodegradation was a primary mechanism for the removal of MEK at the low concentrations of 55 and 430 mg/l, stripping to the atmosphere could become an increasingly important removal mechanism at higher concentrations. TCA did not have any significant removal via biodegradation; its main removal mechanisms were volatilization/stripping.

Polyphosphate Degrading Enzymes

Acinetobacter spp is important in biological phosphate removal from wastewater. The polyphosphate degrading enzymes were studied in *Acinetobacter spp* strain 210A by Groenestun et al.[58] There were three enzymes that can degrade polyphosphate in this strain. These are polyphosphate AMP phosphotransferase (PAMPP), polyphosphatase, and polyphosphatase kinase. PAMPP activity decreased with increasing growth rates of the bacterium. The activity was maximal at a pH of 8.5 and a temperature of 40°C. PAMPP together with adenylate kinases plays the most important role in anaerobic ATP production in *Acinetobacter* strains that accumulate aerobically large amounts of polyphosphate. In samples of activated sludge from different plants, the activity of the adenylate kinase correlated well with the ability of the sludge to remove phosphate biologically from wastewater.

Two Stage Biological/Chemical Treatment of Leachate

Many leachates are characterized by a high content of organic matter (COD, also BOD), high values for ammonia and salts and adsorbable halogenated hydrocarbons (AOX), but fairly moderate levels of heavy metals. With the current restrictions on land use and all the public worry about contaminants from landfill leachate, we need an effective treatment for the leachate. The costs for the transportation of wastes is rising and there is often a need for a pretreatment before it can be discharged at the sewage plant.

This adds to the cost of treatment. Albers and Kayser[59] reported on a two stage treatment process for hazardous waste landfill leachate. They wondered if a combination of chemical oxidation followed by aerobic biological treatment may be an effective means to achieve an extensive purification of these leachates. A research project was initiated in co-operation between the Technical University of Braunschweig and the Nuclear Research Centre in Karlsruhe, W. Germany. The aim of the study was to find a technique that would eliminate the majority of the biodegradable organic compounds in the first biological stage with an activated sludge plant. The biological treatment which consisted of several completely mixed, continuous flow activated sludge reactors with internal sludge recycle was followed by a chemical oxidation step to convert the non-biodegradable matte into more easily biodegraded substances (cracking). Fenten's reagent or ozone were used as oxidizing agents. The organic substances produced in the reaction are degraded by the 124 second biological treatment stage. The reactors were started up with excess sludge from nitrifying activated sludge plants treating sanitary landfill leachates. Orthophosphoric acid was added to the influent to adjust the BOD/P ratio to 100:1. The results showed that the high removal efficiencies were obtained for various leachates tested (COD 80-90%, BOD effluents <25 mg/l). Very low loadings had to be applied and there were some problems with poor activated sludge settling characteristics. Biological nitrification of the high ammoniacal nitrogen contents proved to be a very sensitive process with low reaction rates mainly due to the high salt concentration and some inhibitory factors.

Chemical oxidation with hydrogen peroxide was able to break down parts of the COD still present in the effluents and make them biodegradable, but probably not to such an extent that no further post-treatment steps are needed to meet discharge criteria.

ANAEROBIC BACTERIA

Metabolism

Anaerobic bacteria differ from aerobic bacteria in that they do not use oxygen and are probably harmed by its presence. Research into anaerobic processes of biodegradation have not been studied as extensively as research into aerobic processes. Use of one or the other have concomitant advantages and problems. The aerobic bacteria are easier to grow in the laboratory and generally show a greater metabolic diversity compared to the anaerobes, but also for effective biodegradation they need aeration which is costly, and since they are so metabolically efficient their biomass accumulates and can foul the system. Also they are susceptible to poisoning by toxics, particularly heavy metals. The anaerobic bacterial are more resistant to toxicity and don't accumulate biomass as much, as well as not needing aeration. Also they can work in-situ if they don't have exposure to the atmosphere. However they are difficult to culture in the lab and need specialized handling techniques. They need pre-reduced media to grow on and their growth takes place very slowly, typically from weeks to months. But anaerobic bacteria has a completely different type of metabolism from the aerobic bacteria and therefore present the possibility of a degradation pathway not available to aerobic bacteria.

There are two metabolism patterns with respect to electron acceptors:

- Respiratory--where the terminal electron acceptor is inorganic and supplied by the environment, for example oxygen.
- Fermentative--where the terminal electron acceptor is organic and supplied by the cells own metabolism, for example lactic acid.

A traditional view has been that respiration gives us complete mineralization

$$\text{Glucose} + O_2 \rightarrow CO_2 + H_2O$$

but that fermentation does not result in mineralization

$$\text{Glucose} \rightarrow \text{lactate, acetate, etc.}$$

Some fermentation techniques though çan lead to mineralization, for example oxides of nitrogen can act as terminal electron acceptance with the end result of nitrogen and carbon dioxide. These denitrifying bacteria are almost as energy efficient as aerobic bacteria. The methanogenic bacteria use carbon dioxide as the terminal electron acceptor, producing methane. Similarly, the sulfidogenic bacteria use sulfate ions to accept electrons and produce sulfide. These are respiratory type reactions but all anaerobic. Methanogens and sufidogens though can only give complete mineralization with a sequential operation of organisms called a consortium. Methanogens can oxidize hydrogen, formate, acetate, methanol, and methylamine. Sulfidogens can oxidize hydrogen, formate, acetate, and a few short chain alcohols.

It appears that there are more than one type of organism involved in biodegradation, the primary degraders ferment the organic compound to a typical fermentation product. The methanogens and sulfidogens cannot utilize these products and yet they do not accumulate and inhibit the system. It appears that there are a group of bacteria which use protons as the terminal electron acceptor to produce hydrogen. This reaction has a high redox potential, so that it is unfavorable unless the product has been stripped from the system as the methanogens and sulfidogens do. The consortium then consists of the primary degraders which determine the organic that can be degraded. This is the first step in the procedure and unless the primary degraders can utilize the compound, then no more degradation can occur. Next are the obligate protons who ferment the products from the primary degraders larger than acetate and the electron sink organisms who oxidize hydrogen, formate, and acetate and who are necessary for the obligate protons to survive.

Anaerobic Processes

There are several processes which have been developed for anaerobic degradation. Some of these are: conventional, aerobic contact process, anaerobic upflow filter, expanded bed, sludge blankets, and anaerobic rotating biological contractors. All involve procedures for influent and effluent control, and gas collection. An anaerobic system can have the advantage that a usuable product such as methane is formed.

The **conventional process** consists of a heated reactor containing waste and bacteria. The waste, usually sewage sludge, can be added either continuously or periodically (semi-batch). There is compromise between the hydraulic retention time (HRT) and solid retention time (SRT), the longer the SRT, the longer the bacteria have to degrade the solids.

The **aerobic contact process** or activated sludge process involves two stages. The first reactor is like the conventional model. The digested waste undergoes secondary treatment in the second stage where it is further clarified. Some of the waste is recycled to the first stage. Waste is fed to the reactor that has a high concentration of bacteria and a high concentration of solids. The solids are controlled by solids wastage and the thickening ability of the sludge. Here the HRT can be decreased if there is an adequate HRT.

In an **anaerobic upflow filter**, the bacteria grow on a packed bed that trap a high concentration of solids so that there is a long solid retention time. The high SRT's lead to increased biodegradation at normal temperatures.

In the **anaerobic expanded bed process**, the fluid velocities are increased so that they can fluidize and expand sand beds containing the bacteria so that there is a large surface area in the reactor and hence a large contact area for the bacteria and hence more biodegradation is possible.

The upflow **activated sludge blanket process** consists of a reactor that has two components; a sludge bed which is formed by the settling and thickening abilities of the bacteria so that there is good retention of solids and a sludge blanket above this that consists of highly flocculated sludge formed from the production of gas from the sludge bed which ensures mixing of the bacteria and waste.

An **anaerobic rotating biological contractor** (AnRBC) is a series of disks, mounted on a horizontal shaft in an enclosed cylindrical tank. The disks are partially submerged in wastewater and rotated to facilitate mixing and gas transfer. So it is similar to an aerobic RBC but is kept under anoxic conditions. Since anaerobic bacteria do not accumulate as much biomass as aerobic bacteria clogging is less of a problem and can be decreased if the disk is rotated enough to have a sloughing off of the film. Laquidare et al.[60] studied the operation of an anaerobic biological rotating contractor. They seeded the RBC with anaerobically digested primary sludge and acclimated it with a growth substrate for four weeks. Next they fed the AnRBC reactors with organic load for 211 days. Maximum organic removal was observed at low application rates; methane production of 20 L/M^2/day. Solids were produced so that a polishing system would be needed if surface discharge was needed.

Perchloroethylene

Perchloroethylene (PCE) and Trichloroethylene (TCE) are very common contaminants of groundwater. There is concern about them since they are so common and they are carcinogens. They are usually resistant to aerobic degradation except by methanotrophs which can degrade them to carbon dioxide with no other volatile chlorinated compounds produced in the process as discussed earlier. The anaerobic sequential reductive dechlorination pathway is from PCE>TCE>DCE (dichloroethylene)>VC (vinyl chloride). Fathepure et al.[61] identified bacterial cultures that could biodegrade PCE. They tested several methanogens and found *Methanosarcina mazei* which is very common in methanogenic environments and *Methanosarcina* sp showed PCE degradation but worked best using methanol as the growth substrate. But DCB-1, a strain isolated from a consortium that dechlorinates 3-chlorobenzoate was grown on pyruvate dechlorinates PCE 3-5 times as fast. Also tested were various enrichments grown on chlorophenol, 3-chlorobenzoate, chlorocatechol, and chlorosorcinol as the sole added carbon source. Each enrichment had been growing for at least two years. It was found that DCB-1 was more effective at degradation of PCE and TCE in a cosortium with BZ-2 (a benzoate degrader) and *Methanospirillium* sp strain where PCE was completely reduced in one week. The problem with reductive dechlorination is that it is more effective with a more chlorinated compound, but may not dechlorinate it totally and hence other chlorinated compounds are produced in the process. Methanotrophic oxidation on the other hand is more effective at dechlorinating less chlorinated compounds and doesn't have chlorinated intermediates, but cannot work on PCE. The authors suggest that a two stage biological process may be implemented with reductive dechlorination followed by methanotrophic oxidation and that this may be useful for the decontamination of waters with C1 and C2 chlorinated solvents.

Coal Gasification Wastewater

Another anaerobic process that can be used in conjunction with an aerobic system is for treating coal gasification wastewater. Coal can be converted to a gas so that we use some of the vast coal reserves, but in the process a polluted wastewater is produced. This contains phenols, ammonia, thiocyanate, and other organic and inorganics. Work by Pfeffer et al.[62] describes an anaerobic-aerobic system for biodegradation. It consists of a granular activated carbon anaerobic filter followed by a nitrification activated sludge. The anaerobic filter removes solids as they pass through it. The bacteria are a consortium with fermentators and methanogens. The fermentation products are diluted until the methanogens can utilize them. The addition of granular activated carbon (GAC) improves the process but it was found that it had to be replaced in the experiment otherwise the existing GAC became saturated and toxic compounds built up and inhibited the bacteria. The anaerobic filter was very efficient at removal of phenols. The nitrification activated sludge system was responsible for a polishing of the effluent, reducing both the COD and DOC levels as well as oxidizing ammonia, cyanide, and thiocyanate to nitrates and carbon dioxide. Cyanide, thiocyanate, and ammonia were decreased 89%, 83%, and 95%, respectively. The two stage process decreased COD and DOC by more than 98%.

With the increased efficiencies of the devices developed for anaerobic treatment, the anaerobic process is expected to become a valuable method in wastewater treatment. It has the advantage of low energy consideration, small amounts of sludge generated, simple equipment, and methane recovery.

Tannery Wastes

Yang[63] studied the degradation of tannery wastes which contained organic pollutants with large molecular weights and complex structures that are not easily hydrolyzed. He found that it is feasible to treat tannery wastes with anaerobic bacteria although a long retention time was needed. The sludge used was initially from a sewage plant and needed two months acclimation. Also pH has to be carefully monitored, or the flocculating ability of the bacteria was altered and decreased degradation of the wastewater. He found a 60-77% removal with an influent COD of 8600 mg/l and a volumetric loading of 1.2-2.9 kg $CODm^3$/d. The efficiency was increased by maintaining large amounts of activated sludge. At Moench Tanning, a large operation located in Gowanda, there have been a series of pilot-scale waste treatment studies.[64] One phase of the project used an anaerobic fluid-bed reactor where the bacteria growing on sand particles convert the waste into a mixture of gases with fuel value. The particles are suspended and mixed by the flowing wastewater and allow good contact between the toxic and the bacteria. Problems were encountered because the bacteria were inhibited by high sulfur levels. In the second phase, a sludge bed/fixed film bioreactor was combined with a sulfur stripping process.

1,1,1-Trichloroethane Degradation In-Situ

Another use for anaerobic degradation was demonstrated by Boyer et al.[65] This was a pilot plant study of the degradation of 1,1,1,-trichloroethane (TCA) in-situ. There had been a spill of TCA into an unconfined aquifer. Laboratory work indicated that TCA is assimilated by methanogenic consortia and since disturbing the soil would lead to volatization of TCA, an in-situ remediation process was undertaken. Two packed bed reactors (lysimeters) were prepared with granular activated carbon and nutrients and trace elements added which were necessary for bacterial growth. The reactors were inoculated with a mixture of waste secondary sludge and anaerobic digester sludge. Glucose was added as a secondary carbon source. The reactors were fed the contaminated wastewater from pumping wells but a hurricane and a drought adversely effected the supply and so a pump was installed to maintain a good supply of wastewater to the bacteria. The authors also encountered several problems with the reactors. There was soil matrix damage due to high sodium concentrations which made the clay swell and the cavities to collapse resulting in a decreased rate of biodegradation. This was alleviated by the addition of a suitable cation concentration by adding calcium nitrate. Also there were decreased velocities of treated wastewater which were eventually found to be mechanical in origin and so the reactors were redesigned to overcome this. The ultimate success of the biodegradation was conversion of TCA to 20 ppb, with no other halohydrocarbons found. So although the study encountered several problems it was found that anaerobic degradation in-situ is a viable process for the remediation of soil and groundwater contaminated with TCA.

Patent for Haloaromatic Compounds

The Occidental Chemical Corporation, Niagara Falls, NY obtained a patent (Aug. 1988) for a method to degrade haloaromatic compounds in a soil medium in-situ using *Klebsiella oxytoca* strain SAL-18A and its mutants. These are facultatively anaerobic microorganisms.

2,4-Dichlorophenol

Temperature in anaerobic sediments can widely vary throughout the year and this may have effects on the fate of toxic chemicals. A study by Kehring et al.[66] reported the effect of temperature on the reductive dechlorination of 2,4-dichlorophenoxyacetic acid and 2,4,5-trichlorophenoxyacetic acid. These chemicals were studies since the aerobic degradation of them is slow and they pass into the anaerobic zone. The proposed pathway of the degradation is reductive dechlorination to phenol and then further metabolism to methane and carbon dioxide. The study showed that degradation in unacclaimed sediments dominated by methanogenic bacteria only occurred at temperatures between 5-50°C, although there was some methane still formed to 60°C. The first product was 4CP and at least two bacterial types were involved. 4CP appeared to be the rate limiting step. Storing the sediments at cold temperatures decreased the amount of degradation, probably because the bacteria had decreased and they needed time to grow. Temperatures are important since they effect both the composition of the community and the enzymatic catabolic reaction.

FUNGI

Dioxin

Use of fungi for biodegradation has been limited, but some studies suggest that it may be useful. Dioxin (2,3,7,8-TCDD) is one of the most toxic compounds known to us.

The EPA estimates a generation of dioxin of 1300 tons/year in the U.S. There is also a stored 5000 metric tons waiting for treatment. Dioxin is a very stable compound and biodegrades slowly. It is colorless

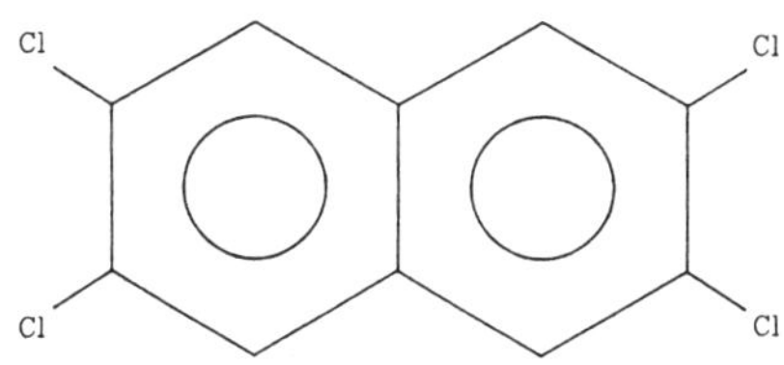

Figure 13. Chemical structure of dioxin.

and crystalline at room temperatures and only slightly soluble in water and most organic liquids. Currently it is usually treated by incineration (mainly liquid injection). Incineration is a viable treatment technique, but it is not without its problems. There has been some recent research[67,68] that suggests that dioxin may be biodegraded by a fungus. The white rot fungus *Phanerochaete chrysosporium* secretes a unique hydrogen peroxide dependent enzyme capable of degrading lignin. It also appears effective in degrading highly toxic and refractory organic pollutants including dioxin and organohalides such as lindane, DDT, 4,5,6-trichlorophenol, and 2,4,6-trichlorophenol. So far the work has been only done in the laboratory but EPA's HWERL (hazardous waste engineering research lab) plans to test the enzyme at contaminated sites in the future. CE Environmental, Roseland, New Jersey, has obtained exclusive rights to a process that uses a white rot fungus to treat hazardous wastes. The system developed by Biotechnology Research

Associates (Raleigh, NC) involves treatment of the waste in tanks. Research so far[69] shows that the process can degrade trichlorophenol, trinitrotoluene, DDT, lindane, PCB, benzopyrene, dioxin, and other chlorinated organics. Work with dioxin degradation is still at the initial stage and needs further testing. Arthur et al.[70] studies the effect of TCDD on soil microorganisms and found slightly less fungi and oligotrophic bacteria than control soils, but there was still a considerable diversity of microorganisms regardless of the concentration of TCDD. So its presence did not appear to exert an influence of soil enzymatic activity. The authors postulate that either TCDD is not toxic to the soil microorganisms or it is not bioavailable (chemical features meant that it adheres to the soil particles). If needed many organisms can grow in its presence then perhaps more than the white rot fungus may be able to degrade it.

PAH Degradation

Bumpus[71] investigated the ability of *P. chrysosporim* to degrade a mixture of PAH since contamination is more likely to consist of more than one chemical and so if the fungus is to be of any practical use it must be able to degrade a mixture. The study used anthracene oil, a distillation product obtained by the fractional distillation of coal tar. At least 22 PAHs, including all the most abundant PAH components present in anthracene oil, underwent 70-100% disappearance during 27 days of incubation with nutrient limited cultures of this fungus. The lignin-degrading enzyme system is expressed under nutrient limiting conditions. These results suggest that the fungus may be useful for the decontamination of sites contaminated with PAHs.

Stephen Aust of Utah State University and others[72] are doing bench scale composting studies using a white rot fungus. Aust says bacteria, the active agents in all other biological systems, each use only one or a few enzymes, and each will degrade only a relatively narrow band of chemicals. The white rot fungus generates a wider range of enzymes, he says, and each will degrade a wider range of organic chemicals. Bacterial enzymes work only within the bacterial cell, and thus will degrade only the dissolved chemicals, while the fungus actively secretes enzymes externally, so will degrade some insoluble chemicals too. The technique has not yet been applied on large scale field trials.

Selenium

Another fungus may be helpful in the decontamination of selenium.[73] In Kesterson National Wildlife Refuge, California selenium had been unnaturally concentrated in farm-drainage water. Originally a system of underground pipes was to have carried the salt-laden runoff from San Joaquin Valley farms into San Francisco Bay but it stopped in Kesterson, and along with the salt, the water carried selenium. The effects were seen in birth deformities of birds feeding on plant seeds which had concentrated selenium. A fungus has been found that converts the selenium into a harmless gas and although only at the research stage it may prove to be a treatment alternative.

Immobilization of Phenolics

A problem with landfills has been the toxic nature of leachate. Remediation techniques have involved moving the soil or treatment of the groundwater, both of which are difficult and costly. A study by Shannon[74] used an enzyme produced by the fungus *Geotrichum candidum* to immobilize phenolic compounds in the soil and hence leave more time for them to be degraded. The enzyme did not immobilize anilines, benz(a)pyrenes which was good; the enzyme would be no good if it encouraged normally stable pollutants to leach. They used 4-methyl phenol in a column as a control group and one treated it with the enzyme. They found that less of it leached out when treated with the enzyme due to its polymerization and incorporative abilities. 2,4-DCP is less reactive and therefore a more difficult candidate to use in the immobilization test. They found that in just one day the amount that leached out was half that of the control group. This illustrates that groundwater pollution by phenolic substitutes could be controlled by their immobilization technique. More search is needed on how to maximize the oxidative enzyme production, perfect its storage and field application, and to demonstrate a lasting decrease in the toxicity of phenolic compounds.

Metalaxyl Degradation

Finally, a fungus has been found able to degrade a fungicide. Fungicides have been used copiously in the past and can be a source of concern if they are persistent in the environment. Although the study did not discern an ultimate fate, or if the fungicide could be mineralized they found the metabolites and isolated possible degradation pathways. Metalaxyl (N-(2,6-dimethylphenyl)N(methoxyacetyl)-(alanine methyl ester)) is a fungicide with residual and systemic activity against plant pathogens. it is persistent in sterilized soil. Zheng et al.[75] concluded that metalaxyl could be degraded by the fungus *Syncephalastrum racemosum* when after incubating for 21 days 80% of the compound was transformed. The transformation decreased as the concentration of metalaxyl increased to 200 ug/ml indicating that the microbial enzyme responsible for degradation was inhibited by higher concentrations of the fungicide. The fungus could not utilize metalaxyl as the sole source of carbon and energy and hence the authors think that cometabolism is necessary. Two isomeric metabolites and a mixture of two other isomeric metabolites were isolated. These were N-3-methyl-6-hydroxymethylphenyl-N; N-2-hydroxy-methyl-7-methylphenyl-N (methoxy acetyl)-alanine methyl ester; N-3-hydroxy and N-5-hydroxy 2,6-dimethyl phenyl-N- (methoxyacetyl) alanine methyl ester. The major transformation mechanisms by the fungus, deduced from the presence of the metabolites are believed to be either hydroxylation on the aromatic ring or benzylic hydroxylation of the methyl side chain of the ring.

CONCLUSIONS

In conclusion, biological degradation using bacteria and some fungal species has a role in the present and future degradation of hazardous waste compounds. Using them is advantageous in that they can utilize xenobiotics, provided some period of acclimation or adaptation is ensured. This period of adaptation may result in new enzymatic pathways, or succession of bacterial groups or genetic changes such as mutations or plasmid exchanges. In addition, the bacteria we often would like to degrade a specific compound can be isolated from the same area or type of area where we find the compound. The bacterial groups appear to be able to degrade a wide range of chemicals, even the four ring polycyclic aromatic hydrocarbons. Problems can appear in the implementation and engineering stages. The microorganisms are sensitive to xenobiotic concentrations, nutrients, temperature, pH, and other organic substrates. Some strains are inhibited by another growth substrate while some are enhanced by its presence.

Each population has specific requirements and so the engineering processes in place need extensive monitoring. How successful the strategies are in-situ also depends in part on the external factors. Biological degradation has also been shown to be successful when used in conjunction with abiotic treatment techniques. There are so many parameters impinging on the biological treatment system that copious research is needed at the laboratory, bench scale, pilot studies, and in-situ studies before we can know how to optimize degradation. We must know what the degradation pathways are, the toxic intermediates if any, and if more stable products are formed. Each situation is unique; a study may show that a compound is recalcitrant but the bacteria may require an extra growth substrate or symbiotic partner to utilize it. We still have unanswered doubts about the analytical techniques that test for recalcitrance and mineralization. But as our understanding of the enzymatic pathways, ecological succession, genetics, and microbiological basis of degradation grows, so will the systems we invent to degrade the xenobiotics.

So in spite of the limitations of using biological control mechanisms, the literature has shown that microorganisms have a large capacity for degrading toxic chemicals.

REFERENCES

1. "State Decision Makers Guide for Hazardous Waste Management," *U.S. EPA*, 1977.

2. de Young, G. "EPA and Industry Push for New Superfund Solutions," *Chemical Week*, Feb. 24, 1988, p. 28.

3. "Waste Minimization Involves Recycling and Source Reduction," *Environ News*, vol. 3 #1, pp. 14-15.

4. Valentino, F.W., and G.E. Walmet. "Industrial Waste Reduction: The Process Problem," *Environment*, Sept. 1986, vol. 28 #7, p. 16(10).

5. Dawson, G. W., and B.S. Mercer. *Hazardous Waste Management*. New York: Wiley & Sons, 1986.

6. Bowers, A.R., W.W. Eckenfelder, P. Gaddipati, and R.M. Monsen. "Toxicity Reduction in Industrial Wastewater Discharge," *Pollution Engineering*, Feb. 1988, vol. 20 #2, pp. 68-73.

7. Grosse, D.W., M. Koczwara, J.E. Park, and R.J. Lesiecki. "Activated Sludge Treatment of Selected Aqueous Organic Hazardous Waste Compounds," *Proceedings of the 42nd Industrial Waste Conference at Purdue University*, May 1987.

8. Valentino, F.W., and G.E. Walmet. "Industrial Waste Reduction: The Process Problem," *Pollution Engineering*, Feb. 1988, vol. 20 #2, pp. 68-73.

9. Turner, R.J. "A Review of Treatment Alternatives for Wastes Containing Non-Solvent Halogenated Organics," *JAPCA*, June 1986, vol. 36 #6.

10. de Young, G. "EPA and Industry Push for New Superfund Solutions," *Chemical Week*, Feb. 14, 1988.

11. Turner, R.J. "A Review of Treatment Alternatives for Wastes Containing Non-Solvent Halogenated Organics," *JAPCA*, June 1986, vol. 36 #6.

12. Ryan, J.R., R. Kabrick, and R. Loehr. "Biological Treatment of Hazardous Wastes," *Civil Engineering*, Feb. 1988, pp. 65-67.

13. Grosse, D.W., M. Koczwara, J.E. Park, and R.J. Lesiecki. "Activated Sludge Treatment of Selected Aqueous Organic Hazardous Waste Compounds," *Proceedings of the 42nd Industrial Waste Conference at Purdue University*, May 1987.

14. Ryan, J.R., R. Kabrick, and R. Loehr. "Biological Treatment of Hazardous Wastes," *Civil Engineering*, Feb. 1988, pp. 65-67.

15. Dawson, G. W., and B.S. Mercer. *Hazardous Waste Management*, New York: Wiley & Sons, 1986.

16. Park, K.S., R.C. Sims, W.J. Doucette, and J.E. Matthews. "Biological Transformation and Detoxification of 7,12-dimethylbenzanthracene in Soil Systems," *J. Water Pollution Control Federation*, Oct. 1988, vol. 60 #10, pp. 1822-1825.

17. Grady, C.P., and J.R. Leslie. "Biodegradation of Hazardous Wastes by Conventional Biological Treatment," *Hazardous Waste and Hazardous Materials*, Nov. 1986, vol. 3, pp. 336-365.

18. Vandenbergh, P.A., R.H. Olsen, and J.H. Colaruoble. "Isolation and Genetic Characterization of Bacteria That Degrades Chloroaromatic Compounds," Oct. 1987, pp. 737-739.

19. Reed, J.P., R.J. Kremer, and A.J. Keastor. "Characterization of Microorganisms in Soils Exhibiting Accelerated Pesticide Degradation," *Bull. Environ. Contam. Toxicol.*, 1987, 39:776-782.

20. Bowers, A.R., W.W. Eckenfelder, P. Gaddipati, and R.M. Monsen. "Toxicity Reduction in Industrial Wastewater Discharge," *Pollution Engineering*, Feb. 1988, vol. 20 #2, pp. 68-73.

21. Ghisalba, O. "Microbial Degradation of Chemical Waste, An Alternative to Physical Methods of Waste Disposal," *Experientia*, 1983, p. 39.

22. Pitter, P. "Determination of Biological Degradability of Organic Substances," *Water Research*, 1976, vol. 10, p. 231(5).

23. Chudoba, J. "Inhibitory Effect of Refractory Organic Compounds Produced by Activated Sludge Microorganisms on Microbial Activity and Flocculation," *Water Research*, 1985, vol. 19 #2, pp. 197-200.

24. Harman, J., C. Lange, and M. Matsumora. "Toxicity Evaluation of Persistent Organic Contaminants," *Proceedings of the 42nd Industrial Waste Conference at Purdue University*, May 1987.

25. Blok, R.L., J. deMorsier, H. Wellens, and W.J. Bontinck. "Evaluation of the Toxicity of Substances to be Assessed for Biodegradability," *Chemosphere*, 1987, vol. 16, pp. 2259-2277.

26. Wilson, D. "Report on Decontamination of PCB Bearing Sediments," *U.S. EPA Cincinnati*, OH, Oct. 1987.

27. de Young, G. "EPA and Industry Push for New Superfund Solutions," *Chemical Week*, Feb. 24, 1988.

28. Kampbell, D.H., J.T. Wilson, H.W. Read, and T. Stocksdale. "Removal of Volatile Aliphatic Hydrocarbons in a Soil Bioreactor," *JAPCA*, Oct. 1987, vol. 37 #10, pp. 1236-1240.

29. Skladany, G.J., J.M. Thomas, G. Fisher, and R. Ramachandran. "The Design, Economics and Operation of Biotransformation System for Ketone Control and Solvent Recovery Process Waters," *Proceedings of the 42nd Industrial Waste Conference at Purdue University*, May 1987.

30. Ely, D.K., and D.A. Heffner. US Patent 465902 Pollution Abstracts, 1988, vol. 19 #6, pp. 88-08439P.

31. Ying, W.C., R.R. Bonk, and S.A. Sojka. "Treatment of a Landfill Leachate in Powdered Activated Carbon Enhanced Sequencing Batch Reactors," *Environmental Progress*, Feb. 1987.

32. Weber, S.A., and M.R. Matsumoto. "Feasibility of Intermittent Biological Treatment for Hazardous Wastes," *Environmental Progress*, Aug. 1987, vol. 6 #3, pp. 166-171.

33. Little, C.D., A.V. Palumbo, S.E. Herbes, M.E. Lidstrom, R.L. Tyndall, and P.J. Gilmer. "Trichloroethylene Biodegradation by a Methane-Oxidating Bacterium," *Applied and Env Mic*, Apr. 1988, p. 951.

34. Heitkemp, M.A., W. Franklin, and C.E. Cerniglia. "Microbial Metabolism of Polycyclic Aromatic Hydrocarbons: Isolation and Characterization of a Pyrene Degrading Bacterium," *Applied and Env Mic*, Oct. 1988, vol. 54 #10, pp. 2549-2559.

35. Heitkemp, M.A., J.P. Freeman, D.W. Miller, and C.E. Cerniglia. "Pyrene Degradation of a *Mycobacterium* sp: Identification of Ring Oxidation and Ring Fission Products," *Applied and Env Mic*, Oct. 1988, vol. 54 #10, pp. 2556-2565.

36. Bauer, J.E., and D.G. Capone. "Effects of Co-Occurring Aromatic Hydrocarbons on the Degradation of Individual Polycyclic Aromatic Hydrocarbons in Marine Sediment Studies," *Applied Env Mic*, July 1988, vol. 54, pp. 1649-1655.

37. Mihelck, J.R., and R.G. Luthy. "Degradation of Polycyclic Aromatic Hydrocarbons Under Various Redox Conditions in Soil and Water Systems," *Applied and Env Mic*, May 1988, vol. 54 #5, pp. 1182-1187.

38. Mahaffey, W.R., D.T. Gibson, and C.E. Cerniglia. "Bacterial Oxidation of Chemical Carcinogens: Formatting of Polycyclic Aromatic Acids from Benz(A) Anthracene," *Applied Env Mic*, Oct. 1988, vol. 54 #10, pp. 2415-2423.

39. Bauer, J.E., R.P. Kerr, M.F. Baushsta, C.J. Decker, and D.G. Capone. "Stimulation of Microbial Activities and Polycyclic Aromatic Hydrocarbons Degraded in Marine Sediments Inhabited by *Capitella capitata*," *Marine Environ Res*, 1988, vol. 25 #1, pp. 63-84.

40. Guerin, W.F., and G.E. Jones. "Two Stage Mineralization of Phenanthrene by Estuarine Enrichment Cultures," *Applied Env Mic*, Apr. 1988, vol. 54 #4, pp. 929-936.

41. Dieberg, F.E., N.A. Goulet, and T. Debusk. "Removal of Two Chlorinated Compounds from Secondary Domestic Effluent by a Thin Film Technique," *Journal Environmental Quality*, 1987, #4, pp. 321-324.

42. Pelon, W., and J.A. Mayo. "Growth of Mixed Microbial Populations in Ground Water Containing Highly Chlorinated Organic Wastes," *Bull. Environ. Contam. Toxicol.*, Apr. 1987, vol. 38 #4, pp. 661-666.

43. Yaldi, Y., Y. Murakami, and M. Alexander. "Limiting Success of Inoculum to Enhance Biodegradation of Low Concentrations of Organic Chemicals," *Env Sci Tech*, 1988, vol. 22 #12.

44. Renganathen, V. "Possible Involvement of Toluene 2,3-Dioxygenase in Defluorination of 3-Fluor Substituted Benzenes by Toluene Degrading Pseudomonal Species Strain T-12," *Applied Env Mic*, Feb. 1989, pp. 330-334.

45. Topp, E., R.L. Crawford, and R. Hanson. "Influence of Readily Metabolizable Carbon on Pentachlorophenol Metabolism by a Pentachlorophenol Degrading *Flavobacterium* Species," *Applied Env Mic*, Oct. 1988, vol. 54 #10, pp. 2452-2459.

46. Brown, L.R. "Oil Degrading Microorganisms," *Chem Eng Prog*, Oct. 1987, vol. 83 #10, pp. 35-40.

47. Feld, S., N. Wojtonowitz, and R. Marks. "Biodegradation of Oil Field Production Bit Sludges," *Proceedings of the 42nd Industrial Waste Conference at Purdue University*, May 1987.

48. Foght, J.M., D.L. Gutnick, and D.W.S. Westlake. "Effect of Emulsan on the Biodegradation of Crude Oil by Pure and Mixed Bacterial Cultures," *Applied Env Mic*, Jan. 1989, vol. 55 #1, pp. 36-43.

49. Chakrabarty, A. "Microbes Could Break Down Dioxin," *New Scientist*, Feb. 25, 1988, 117:36.

50. Bachman A., P. Walet, P. Wijnan, W. de Bruin, J.L.M. Huntjens, W. Roelofsen, and A.J.B. Zehnder. "Biodegradation of Alpha and Beta Hexachlorocyclohexane in a Soil Slurry Under Different Redox Conditions," *Applied and Env Mic*, Jan. 1988, vol. 54 #1, pp. 143-149.

51. de Bruin,. A.B., J.C. Rienlet, H.H.N. Rynaarb, and A.B. Zehnder. "Aerobic Mineralization of Alpha HCH in Contaminated Soil," *Applied and Env Mic*, Feb. 1988, vol. 54, pp. 548-554.

52. Bergman, A., G. Blomlevist, and U. Oem. "Microbial Degradation by Autochthonous Soil Population of Alpha and Beta HCH Under Anaerobic Field Conditions in Temperature Zones," *Chemosphere*, 1988, vol. 17 #2, pp. 481-487.

53. Konopka, A., D. Knight, and R.F. Turco. "Characterization of a *Pseudomonas* sp Capable of Aniline Degradation in the Presence of a Secondary Carbon Source," *Applied and Env Mic*, Feb. 1989, vol. 55 #2, pp. 383-389.

54. Kanagwa, T., and E. Mikami. "Removal of Methanodiol, Dimethyl Sulfide, Dimethyl Disulfide and Hydrogen Sulfide from Contaminated Air by *Thiobacillus thioparus*," *Applied Env Mic*, Mar. 1989, vol., 55 #3, pp. 555-558.

55. Zheng, S.Y.L., Z. Zheng, R. Zhang, and J.M. Bollag. "Sorption and Metabolism of Metolachlor by a Bacterial Community," *Applied and Env Mic*, Mar. 1989, vol. 55 #3, pp. 733-740.

56. Lee, S.E., H.S. Shin, and B.C. Paik. "Treatment of Cr(vi)-Containing Wastewater by Addition of Powdered Activated Carbon to the Activated Sludge Process," *Water Research*, 1989, vol. 23 #1, pp. 67-72.

57. Grosse, D.W., M. Koczwara, J.E. Park, and R.J. Lesiecki. "Activated Sludge Treatment of Selected Aqueous Organic Hazardous Waste Compounds," *Proceedings of the 42nd Industrial Waste Conference at Purdue University*, May 1987.

58. Groenestun, J.W., M.M.A. Bentvelsen, M.H. Deinema, and A.J.B. Zehnder. "Polyphosphate Degrading Enzymes in *Acinetobacter* spp and Activated Sludge," *Applied and Env Mic*, Jan. 1989, vol. 55 #1, pp. 219-223.

59. Albers, H., and R. Kayser. "Landfill Leachate and Mine Drainage: Two Stage Biological/Chemical Treatment of Hazardous Waste Landfill Leachate," *Proceedings of the 42nd Industrial Waste Conference at Purdue University*, May 1987.

60. Laquidare, M.J., F.C. Blanc, and J.C. O'Shaugnessy. "Development of Biofilm, Operating Characteristics and Operation Control in the Anaerobic Rotation Biological Contractor Process," *58th Annual Conference of the Proceedings of the Industrial Waste Symposium*, 1985.

61. Fathepure, B., J.P. Nengu, and S. Boyd. "Anaerobic Bacteria that Dechlorinate Perchlorethene," *Applied and Env Mic*, Nov. 1987, pp. 2671-2674.

62. Pfeffer, J., and M.T. Suidan. "Anaerobic-Aerobic Process for Treating Coal Gasification Wastewater," *58th Annual Conference of Proceedings of the Industrial Waste Symposium*, 1985.

63. Yang, H. "The Anaerobic Treatment of Tannery Wastewater," *Proceedings of the 42nd Industrial Waste Conference at Purdue University*, May 1987.

64. Valentino, W., and G.E. Walmet. "Industrial Waste Reduction: The Process Problem," *Environment*, Sep. 28, 1986.

65. Boyer, J.D., R.C. Ahlert, and D.S. Kosson. "Pilot Plant Demonstration of In-Situ Biodegradation of 1,1,1-trichloroethane," *J. Water Pollution Control Federation*, Oct. 1988, vol. 60 #10, pp. 1843-1849.

66. Kehring, G.W., J.E. Rogers, and J. Negel. "Anaerobic Biodegradation of 2,4-Dichlorophenol in Freshwater Lake Sediments at Different Temperatures," *Applied Env Mic*, Feb. 1989, vol. 55 #2, pp. 348-353.

67. Freeman, H.H., and R.A. Olexsey. "Treatment Technologies for Hazardous Wastes: Part 1. A Review of Treatment Alternatives for Dioxin Wastes," *JAPCA*, Jan. 1986, vol. 36, pp. 67-75.

68. Olexsey, R.A., and H.M. Freeman. "Managing Dioxin," *Civil Engineering*, Feb. 1987.

69. de Young, G. "EPA and Industry Push for New Superfund Solutions," *Chemical Week*, Feb. 24, 1988.

70. Arthur, M.F., and J.I. Frier. "Microbial Activity in Soils Contaminated with 2,3,7,8-TCDD," *Env Tox Chem*, 1988, vol. 7 #1, pp. 5-13.

71. Bumpus, J. "Biodegradation of Polycyclic romatic Hydrocarbons by *Phanerochaete chrysosporium*," *Applied and Env Mic*, Jan. 1989, vol. 55 #1, pp. 154-158.

72. Ryan, J.R., and R. Kabrick. "Biological Treatment of Hazardous Wastes," *Civil Engineering*, Feb. 1988, pp. 65-67.

73. Budiansky, S., and R. Clemings. "Toxic Wastes: A Little Fungus May Help," *US News & World Report*, Nov. 9, 1987, vol. 103 #19, p. 85(1).

74. Shannon, M.J.R., and R. Bartha. "Immobilization of Leachable Toxic Soil Pollutants by Using Oxidative Enzymes," *Applied Env Mic*, July 1988, vol. 54, pp. 1719-1723.

75. Zheng, Z., S. Liu, A. Freyer, and J. Bollag. "Transformation of Metalaxyl by the Fungus *Synephalastrum racemosum*," *Applied and Env Mic*, Jan. 1989, vol. 55, pp. 66-71.

3 BIOLOGICAL TREATMENT OF INDUSTRIAL WASTES: MUTANT BACTERIA

BIOLOGICAL TREATMENT - OVERVIEW

The objectives of biological treatment of wastewater are to remove the non-settleable colloidal solids and to degrade other organic matter. For industrial wastewater, the objective is to remove or reduce the concentration of organic and inorganic compounds. Because many of these compounds are toxic to microorganisms, pretreatment may be required.

With proper analysis and environmental control, almost all wastewaters can be treated biologically. The removal of carbonaceous BOD is accomplished biologically using a variety of microorganisms, mainly bacteria. The microorganisms are used to convert the organic matter (colloidal and dissolved) into various gases and into cell tissue. Unless the cell tissue that is produced from the organic matter is removed from this solution, treatment has not been accomplished. The only treatment that has been achieved is the conversion of a portion of the organics by bacteria to gaseous end products.

There are three major groups of biological processes:

- Aerobic.
- Anaerobic.
- Combination of aerobic and anaerobic.

The processes are further subdivided into suspended or attached growth systems. The principal application of the above processes are for:

- Removal of carbonaceous organic material (BOD).
- Nitrification.

- Denitrification.
- Stabilization.
- Phosphorous removal.

MICROBIOLOGY BACKGROUND

Microorganisms can be defined as those organisms that are too small to be visible without the aid of a microscope or exist as individual cells. In general, protozoa, algae, fungi, bacteria, and viruses are included in this broad grouping. However, viruses (the smallest) are not cells and differ in many respects from the other organisms. The recent trend is to group microorganisms in three kingdoms: protista, plants, and animals.

In each kingdom the cell is the basic unit of life regardless of the complexity. Each cell contains nucleic acids, hereditary material vital to reproduction. The cytoplasm contains ribonucleic acid (RNA), whose major role is synthesis of proteins. The area of the nucleus is rich in deoxyribonucleic acid (DNA), which contains all the information necessary for the reproduction of all the cell components.[1,2,3] Figure 1 illustrates some of the typical microorganisms that are important to the biological process.

Energy and Carbon Sources

To reproduce and function properly an organism must have a source of energy and carbon for synthesis of new cellular material. Inorganic elements, such as nitrogen and phosphorus, and other trace elements such as sulfur, K, Ca, and Mg are also vital to cell synthesis. Two common sources of cell carbon for microorganisms are CO_2 and organic matter. If an organism derives its carbon from CO_2, it is called *autotrophic*; if it uses organic carbon, *heterotrophic*.

Energy is also needed in synthesis of new cellular material. For autotrophic organisms, the energy can be supplied by the sun (photosynthesis) or by an inorganic redox reaction. If supplied by the sun, the organism's energy is supplied by oxidation or fermentation of organic matter.

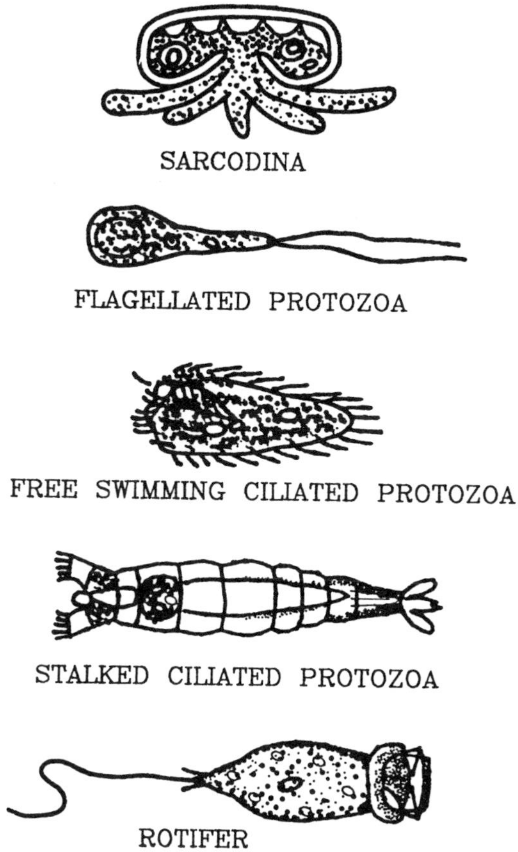

Figure 1. Typical microorganisms that are important in the biological process.

Organisms can also be classified according to ability to use oxygen.

- Aerobic - exists where there is supply of oxygen.
- Anaerobic - exists where there is no oxygen.
- Facultative - can survive with or without oxygen.

Type of Organisms

The following microorganisms are important to the biological treatment processes: bacteria, fungi, algae, protozoa rotifers, crustaceans, and viruses.

Bacteria (procaryotes)--are single cells, and when viewed under a light microscope, it is unusual to see any distinguishable cellular structures. Some structural features can be observed with an electron microscope or sometimes by staining. The cells are small (tenfold difference in average size of procaryotic and eucaryotic microorganisms). Representative size is 0.5 to 1 micron in diameter for the spherical bacteria. (Bacteria fall into three categories: spherical, cylindrical, and helical).

The essential components of a typical bacterial cell are a cell wall, cytoplasmic membrane, a single molecule of DNA, ribosomes, and the cytoplasm. The distinguishing characteristic of procaryotic cell walls is the presence of a mucopeptide layer. This material forms the rigid layer of all bacterial cell walls, including the blue green algae, but has not been found in eucaryotic cells.

Bacteria are about 80% water, 20% dry material of which 90% is organic and 10% inorganic. An approximate formula for the organic portion is $C_5H_7O_2N$. Temperature and pH play a vital role in their life. According to the temperature in which they function, bacteria may be classified as psychrophilic (to 30°C), mesophilic (20-45°C), and thermophilic (45 to 75°C). Most organisms cannot tolerate pH levels above 9.5 or below 4.

Optimum pH for growth is between 6.5 to 7.5. Bacteria can be classified as heterotrophic or autotrophic. In a biological wastewater treatment system, the heterotrophic are important because of their need for organic compounds for carbon for their cells.

Fungi--are the most structurally uniform group of the eucaryotes. The predominant form of growth is filaments (hyphae) which collectively form the filamentous mass called the mycelium. However, yeast are not filamentous fungi that reproduce by budding. The others can reproduce by fission or spore formation.

Geotrichum candidum is the most prevalent fungi in trickling filter slimes (filamentous). It has also been implicated in contributing to sludge bulking in activated sludge units. The ability of fungi to survive under low pH (range of 2 to 9) and low nitrogen conditions makes them important in the biological treatment of industrial waste.

Algae--Many thousands of species of algae exist. They range from unicellular to large aggregates of filamentous cells. Algae are primarily aquatic organisms. They are autotrophic, photosynthetic protists.

Algae produce bad tastes and odors in water supplies and shorten filter runs in filtration plants. In an aerobic facultative oxidation pond, algae are needed to supply oxygen to aerobic heterotrophic bacteria. Algae, like other microorganisms, require inorganic nutrients to reproduce; principally nitrogen and phosphorus. To prevent excessive algae growth in natural waters, emphasis has centered around the removal of nitrogen and/or phosphorus in treatment plant effluents.[32,43]

Protozoa--are mostly single cell organisms, although very complex and highly organized. They are mostly aerobic heterotrophs. They are classified by life cycle and means of locomotion--flagella, cilia, or pseudopodia (amoeba). Protozoa are important both as disease causing organisms and as a vital link in the food chain from bacteria on up. They feed on bacteria and in turn are food for larger organisms. In effect, protozoa act as polishers of the effluent from biological waste treatment processes by consuming bacteria and particulate matter.

Rotifers--are aerobic heterotrophic, multicellular animals. They consume dispersed flocculent bacteria and small particles of organic matter. In waste treatment plant effluents, they indicate an efficient aerobic purification process.

Crustaceans--are also aerobic, heterotrophic, and multicellular animals. They have a hard body or shell. They do not really exist in biotreatment systems; they are, however, normal occupants in natural waters.

Viruses--are small; they can be seen only with an electron microscope. They are parasites and can reproduce only within a living cell. In an infected cell, the metabolism of the cell is used for the manufacture of viruses. Outside the cell, the viruses are inert. They are composed essentially of protein and nucleic acid. Viruses are usually classified by the host they infect: plant, animal, or bacteria, and are generally named for the disease they cause. Viruses are now the most important pathogens to man, since bacterial disease can be treated by antibiotics. Viral infections cannot. And many are found in the feces of man. Therefore, it is necessary to ensure that viruses are controlled in the effluent from waste treatment plants.

BACTERIAL GROWTH

Factors Affecting Growth

Both physical and chemical characteristics of an environment influence microbial growth. These factors can determine the types of organisms that can grow and influence the rate of growth under these conditions.

Temperature

One of the most important physical factors affecting growth is temperature. Each microorganism is able to grow within a specific temperature range. While single species can grow only over a 40°C range, others can grow below 0°C to above 90°C. Based on optimum growth temperatures, microorganisms can be classified as: psychrophiles (less than 20°C), mesophiles (20-45°C), and thermophiles (greater than 45°C.[1,3,43]

pH

Another factor that influences growth rate is pH. Some general statements can be made about microorganisms' pH preferences:

1. Bacteria have optimum pH near 7,(5-9 range).

2. Fungi prefer acid environment (pH minimum of 1 to 3 pH, optimum near 5).

3. Blue-green algae, pH higher than 7.

4. Most protozoa, pH range 5 to 8.

In the treatment of industrial wastes, it is necessary at least to provide initial adjustment of pH if biological treatment is to be successful. In anaerobic treatment of industrial waste, pH control is of greater concern than in aerobic because of a narrow range of pH tolerance.[1,4,26]

Oxygen

Many bacteria can grow only in the absence of oxygen, while many bacteria and fungi and protozoa are capable of growth in either the presence or absence of oxygen. Algae are aerobic organisms.

Oxygen is required for two purposes by aerobes. Mainly, for the electron transport system necessary for generation of energy, and a small amount is used in enzymatic reactions.

Nutrients

Only four elements C, O, N, and H make up 90% of the dry weight of a cell. These elements, plus P and S, comprise the large molecules of the cell. The remainder includes a large number of elements including: K, Na, Ca, Mg, Cl, Fe, etc. Of the four elements comprising the bulk of the cell, i.e., CONH, only C and N are of selected importance. The H and O are derived from water and/or from other compounds used by the cell. Therefore, the major difference in nutritional requirements of microorganisms is the different source of C and N that they can use for synthesis of cellular material.

KINETICS OF GROWTH

Growth Curve

The growth curve shows "the rise and fall" of microbial populations. It is a generalized description of experimental observations. The number of viable organisms are reported as a function of time. The growth pattern, based on number of cells or microbial population in mg/l (X), has several distinct phases:

- Lag.
- Log increasing.
- Log decreasing.
- Stationary.
- Accelerating autodigestion (log increasing).
- Decelerating autodigestion (log decreasing).

These phases can be described as follows:

- **Phase 1**--Time of adjustment to a new environment.
- **Phase 2 and 3**--Exponential and declining growth. Cells generate at a rate determined by ability to process food, where the growth rate equals μ x for the log increasing portion of the curve. Here the growth rate is proportional to the concentration x, and μ is called the specific growth rate. The biomass is increasing exponentially - first order, increasing rate.
- **Phase 4**--Growth ceases. This most likely occurs because of exhaustion of an essential nutrient, and in most biological processes the aim is to make the carbon source the limiting nutrient for the growth of new cells.
- **Phases 5 and 6**--Autodigestion in endogenous phase. Here the death rate exceeds the production of new cells. Microorganisms are forced to metabolize their own protoplasm.

Cultures

In a batch culture, growth in the log increasing phase is proportional to the mass of the bacteria

$$\frac{dX}{dt} = \mu x = r_g$$

where x = concentration of microorganisms

$$\frac{dX}{dt} = r_g = \text{rate of increase in X or rate of bacterial growth}$$

Growth ceases (does not continue indefinitely) when a necessary requirement for growth in the environment is missing: including exhaustion of nutrients, depletion of dissolved oxygen, changes in the chemical environment and production of toxic substances. In wastewater treatment, conditions are controlled so that exhaustion of the carbon supply limits its growth (substrate limiting).

In a continuous culture, growth is limited, and the effect of limiting the substrate or nutrients can be described by the expression proposed by Monod.[1,3]

$$\mu = \frac{\mu_m S}{K_s + S}$$

where μ = specific growth rate, $time^{-1}$

μ_m = max specific growth rate, $time^{-1}$

S = concentration of substrate (growth limiting) in solution, mass/vol

K_s = saturating constant or half velocity constant (concentration of substrate at one/half μ_m), mass/vol

This is an equation of a rectangular hyperbola which fits the data from Monod's experiment, i.e., μ vs. S.[1]

This shape curve is most frequently observed for heterogeneous populations. Although the use of S_o, initial concentration, instead of S is a more accurate representation of this type data, S can be used as an approximate S_o in the equation of the curve.

Substrate Utilization

A portion of the substrate is utilized to create new cells, and the rest is oxidized to inorganic and organic end products. Results of many experiments indicate that the mass of cells produced per unit substrate removed (Y - the cell yield) is constant for pure cultures and also for heterogeneous populations. This value is usually measured at the end of substrate removal and is called the true cell yield. Any Y measured in the autodigestive phase will be lower and is not a value yield measurement. Many factors affect the numerical value of Y; however, the cell yield exhibits constancy for a given species grown on a given carbon source. Therefore, the following relationship has been developed:

$$r_g = Y r_{su}$$

where r_g = rate of bacterial growth

Y = yield coefficient

r_{su} = substrate utilization rate

and r_{su} is also equal to $-\dfrac{kXs}{K_s + S}$, where $k = \dfrac{\mu m}{Y}$, by

combination of the preceding relationships.

In bacterial systems for wastewater treatment, not all cells are in the log growth phase. Cell maintenance and death must be accounted for in the kinetic expressions. The decrease in cell mass due to these factors is assumed proportional to existing cell mass X. Therefore, an endogenous decay term is established:

$$r \text{ (endogenous decay)} = -k_d X$$

When combining terms, the following is the result:

$$r'_g = \frac{\mu m X S}{K_s + S} - k_d X$$

$$r_g = -Y\, r_{su} - k_d X$$

r'_g = actual rate of bacterial growth

A new term is introduced called the net specific growth rate, which is the actual rate of bacterial growth per mass of organisms:

μ' = r_g / X and is equal to

$$\mu' = \frac{\mu m S}{K_s + S} - kd \qquad \text{or}$$

$$\mu' = \frac{YkS}{K_s + S} - kd$$

The effect of endoneous respiration can be incorporated by defining a term called observed yield, Y_{obs}.

$$Y_{obs} = -\frac{r_g}{r_{su}} \text{ or}$$

$$r_g = -Y_{obs}r_{su} \text{ and from above}$$

$$r'_g = -Yr_{su} - k_dX$$

Therefore, $-Y_{obs}\,r_{su} = -Yr_{su} - k_dX$

Hence, Y_{obs} is less than Y (measured in log phase) because of the effect of k_d and is equal to

$$Y - \frac{k_d\,(K_s + S)}{kS}$$

Continuous Treatment

In a continuous process, such as aerobic treatment in a *continuous stirred tank reactor* (CSTR) activated sludge process, a mass balance of organisms, without recycle, and steady state conditions, yields the following relationship:[3]

$$\frac{Q}{V} = \frac{1}{\theta} = \frac{\mu mS}{K_s + S} - k_d$$

here $\frac{V}{Q}$ = hydraulic residence time

V = reactor volume

Q = flowrate

This expression can be written (as derived from previous equations) as:

$$\frac{Q}{V} = \frac{1}{\theta} = \frac{Yr_{su}}{X} - k_d \text{ and is equal to } \mu'$$

The reciprocal of μ' is defined as the mean cell residence time, θ_c

$$\theta_c = \frac{1}{\mu'} \quad \text{or} \quad \mu' = \frac{1}{\theta_c}$$

and $$\theta_c = \frac{\text{mass of cells in reactor}}{\text{mass of cells wasted}}$$

For a system without recycle:

$$\theta_c = \frac{VX}{QX} = \theta \text{ or similarly from above } \frac{1}{\theta} = \mu = \frac{1}{\theta_c}$$

so that the mean cell residence time equals the hydraulic residence time. If a term U (specific substrate utilization rate) is defined as:

$$U = \frac{-r_{su}}{X} = \frac{kS}{K_s+S}$$, then the above equation can be

reduced to the familiar

$$\mu' = \frac{1}{\theta_c} = YU - k_d$$ where U and θ_c are directly related.

For a system with recycle, a similar equation is developed. However, θ_c and U are independent of the hydraulic residence time. Thus, even though the effluent waste concentrations is still related to θ_c or U, θ does not equal θ_c, and θ_c can be altered independent of the system residence time.[34] Table 1 provides a summary of aerobic digestion design parameters.

INDUSTRIAL WASTE TREATMENT PROCESSES

Industrial wastewater is characterized by its large volume, high temperature, high concentration of biodegradable organic matter and suspended solids, high alkalinity or acidity, and by variations of flow. The degree of treatment varies according to the means of disposal, which may be to a municipal sewer system, a receiving body of water, such as a stream, an estuary, or a large body of fresh water, or recovery for reuse. General wastewater treatment processes will probably include:

equalization tanks, settling tanks, precipitation, coagulation, or flocculation, neutralization, and biological processes. Available biological treatment processes include: activated sludge, trickling filter, land disposal, aerobic digestion, stabilization, aerated lagoons, and variations of these such as:

- Rotating Biological Contactors (RBC).
- Deep Shaft.
- Pure Oxygen.
- Activated Carbon.
- Mutant Seeding.

And anaerobic processes are also becoming important particularly for strong organic effluents.

The selection of a biological treatment system will depend upon factors as waste strength, land availability, supply of available technicians, and quality of effluent required, and cost. All of the aforementioned processes should be considered. No two wastes are completely alike, and each industrial wastewater problem must be considered according to its particular situation.

Aerated Processes

Activated sludge remains the predominant biological treatment system for industrial wastes. With proper detention, activated sludge systems perform well for BOD removal. With acclimation well established, removal of some priority pollutants is possible; a properly designed system can even degrade priority pollutants. However, when alcohols and ketones predominate, there is the tendency to form filamentous organisms. Although removals are good, control of settling and recycle is difficult.[27]

Activated sludge, trickling filter, and aerated lagoon treatment typically process wastewaters with less than 1% solids. Waste stabilization ponds process wastewater with less than 0.1% solids. Activated sludge treatment and trickling filter rely on biomass retaining their slimy coating (so they settle better) and they are therefore best operated in the log growth phase. Aerated lagoons and waste stabilization ponds have long retention periods and exhibit endogenous growth.[4]

TABLE 1

AEROBIC DIGESTION DESIGN PARAMETERS

Parameter	Value	Remarks
Detention Time, days	15-20	Waste activated sludge alone
	20-25	Primary + Waste activated sludge
Air Requirements diffuser system, cfm/1000 cu. ft.	20-35[1]	Enough to keep the solids in suspension and maintain a D.O. between 1-2 mg/l.
cfm/1000 cu. ft.	>90[2]	
mechanical system, gp/1000 cu. ft.	1.0-1.25	This level is governed by mixing requirements. Most mechanical aerators in aerobic digesters require bottom mixers for solids concentration greater than 8000 mg/l, especially if deep tanks (>12 feet) are used.
mg O_2/1/hr./1000 mg/l MLSS	3.0	
Minimum Dissolved Oxygen, mg/l	1.0	
Temperature, °C	>15	If sludge temperatures are lower than 15°C, additional detention time should be provided so that stabilization will occur at the lower biological reaction rates.
Volatile Solids Reduction, percent	40-50	
Tank Design		Aerobic digestion tanks are open and generally require no special heat transfer equipment or insulation. For small treatment systems (0.1 mgd), the tank design should be flexible enough so that the digester tank can also act as a sludge thickening unit. If thickening is to be utilized in the aeration tank, sock-type diffusers should be used to minimize clogging.
Power Costs $/yr./lb. BOD removed	2.18	These cost data are based upon three operational plants in Pennsylvania.
$/yr./capita	0.37	

[1]Waste activated sludge alone.
[2]Primary and waste activated sludge.

In general, biological treatment processes are probably the most cost effective techniques for treating aqueous waste streams containing organic contaminants. Certain inorganic compounds may also be treated using microorganisms (although applications are limited); ammonia can be converted to nitrite and nitrate and then nitrogen gas through a controlled sequence of aerobic and anaerobic biological treatment steps.[42] Phosphorus can be concentrated in microorganisms under aerobic conditions and then released under anaerobic conditions.[31] Various post aeration devices are illustrated in Figure 2. The schematic for a continuous-flow aeration unit is shown in Figure 3. Also, a typical aerobic circular digester is illustrated in Figure 4.

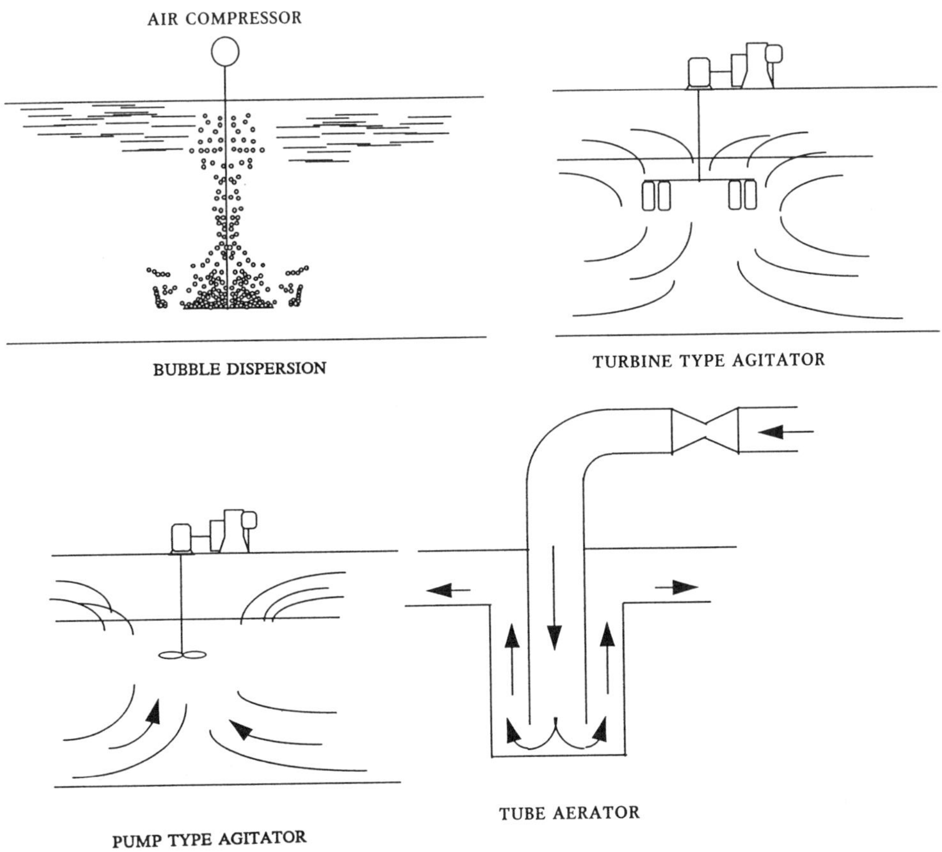

Figure 2. Various post aeration devices.

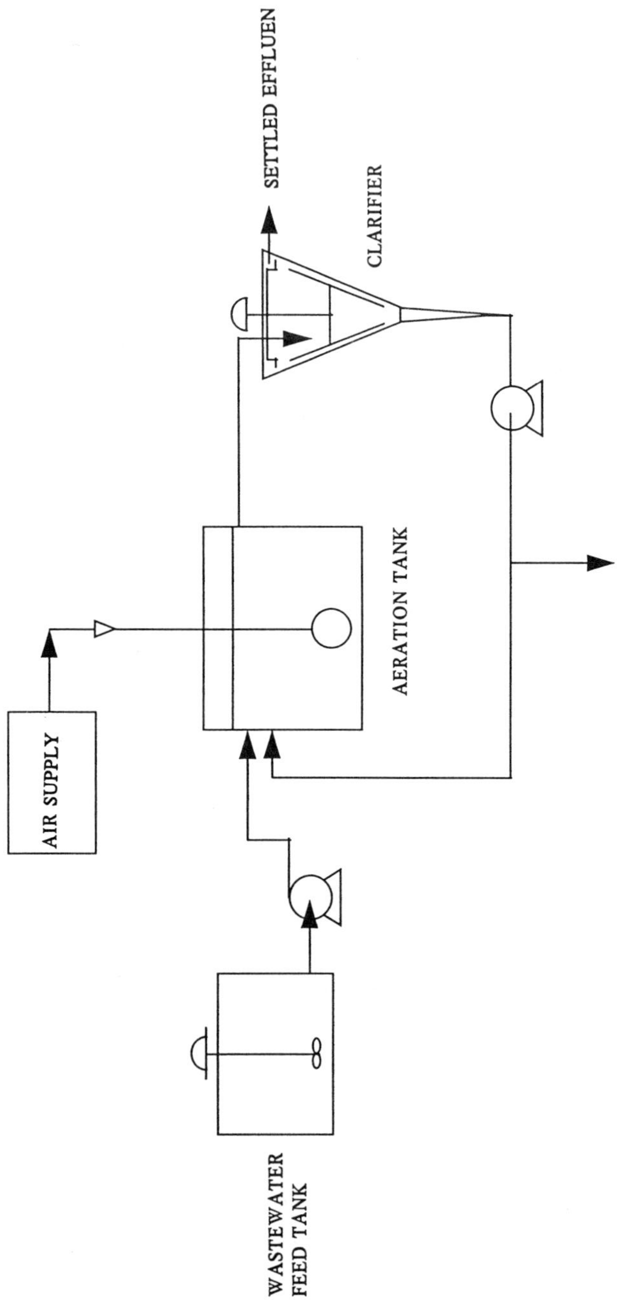

Figure 3. Schematic of a continuous-flow aeration unit.

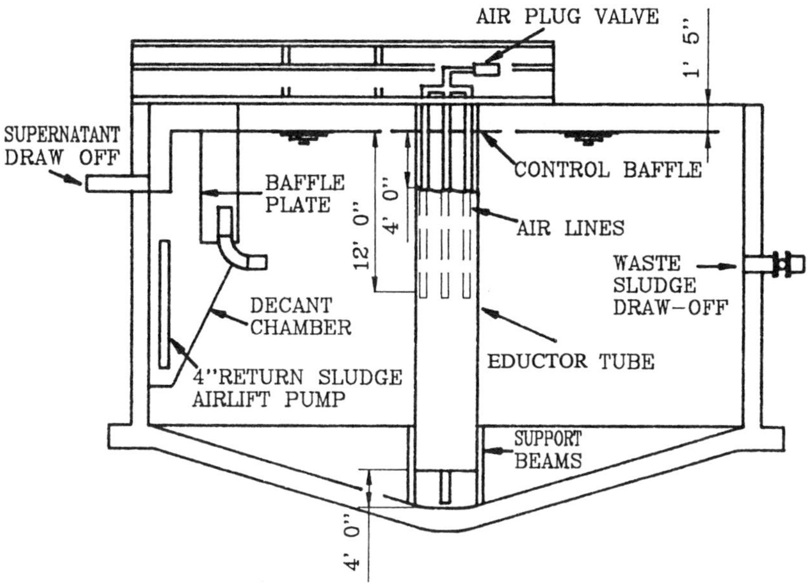

Figure 4. Typical aerobic circular digester.

Activated Sludge (Suspended Growth)

Organic wastes are introduced to a reactor (either plug flow or completely mixed) where the bacterial culture is held in suspension. Aerobic environment is achieved through diffused or mechanical aeration. After a certain time, the biological solids form a floc and settle out and separate from the treated water. A portion of the settled cells is recycled to maintain desired concentration of organisms in the reactor and the rest wasted. Level of biological mass in the reactor depends on desired efficiency and kinetics.[3,4]

As mentioned earlier, a satisfactory floc is necessary for effective separation of the biological solids. The floc formation is promoted by operation at a cell retention time where slime layers exist. Also, important for settling are proper design of a settling unit and proper operation to prevent the presence of filamentous organisms and fungi.

Bacteria in activated sludge are capable of performing hydrolysis and oxidation reactions. The oxidations can be represented by:

Oxidation (dissimilatory)

$$\underset{\text{organics}}{COHNS} + O_2, \xrightarrow{\text{bacteria}} CO_2 \quad NH_3 + \text{energy}$$

Synthesis (assimilatory)

$$COHNS + O_2 + \text{energy} \xrightarrow{\text{bacteria}} C_3H_7NO_2 \text{ (new bacterial cells)}$$

Endogenous Respiration (autooxidation)

$$C_5H_7NO_2 + 5O_2 \rightarrow 5\,CO_2 + 2H_2O + \text{energy (cell tissue)}$$

Both the oxidation of complex hydrocarbons and the hydrolysis of polysaccharides occur outside the cell, and are catalyzed by exoenzymes secreted from the cell wall into the surrounding aqueous environment. Oxidation is conducted by aerobic organisms which use dissolved oxygen present in the biological system. Hydrolytic reactions are caused by aerobic organisms using water present in the biological system.

The activated sludge process was developed in England by the turn of the century. Many versions of the original process are used today, but fundamentally they are all similar.

Process modifications of conventional activated sludge include the following:

- **Contact stabilization**--involves aerating activated sludge on its return trip to the aeration tank so that sorbed organics are decomposed.
- **Step aeration**--involves admitting influent wastewater at multiple points along the aeration tank.
- **Extended aeration**--used where there are low organic loadings and it is desirable to minimize sludge residue. Involves longer aeration retention so that endogenous respiration of the biomass is achieved.
- **Pure oxygen**--used where there are high organic and trace metals concentrations to maintain a high dissolved oxygen level and a high biomass concentration. Involves closed

aeration tank with mechanical mixers receiving wastewater and oxygen gas.

Pure oxygen systems have been shown to operate effectively at MLVSS concentrations of 4000 to 8000 mg/l while conventional air activated sludge systems typically operate at 1500 to 2500 mg/l MLVSS.[13]

- **Activated carbon**--Carbon is sometimes added to activated sludge systems to provide adsorptive capacity.[13] The process combines aerobic treatment with activated carbon to destroy materials refractory to conventional treatment. Dupont has developed a process called PACT (powdered activated carbon treatment) for treatment of broad types of aqueous waste streams. In the PACT process, powdered carbon is added to the aeration basin of an activated sludge system and is capable of capturing and removing non-biodegradable organic compounds. Dupont has such a treatment plant (40 mgd) at its Chambers Works in Deepwater, New Jersey.[45]
- **Deep shaft**--A vertical underground shaft is used to provide highly efficient transfer of oxygen for biodegradation of sewage and organic industrial effluents. Depending on the effluent composition and flow, shaft dimensions can be 50 - 150 meters deep and 0.5 to 10 meters in diameter. Influent is fed into the top of the downflow section of the shaft and mixed with air injected from a standard compressor. Process advantages include a high intensity of oxygen uptake, high utilization of transferred oxygen, and high energy economy. The low land area required and minimal environmental impact make the deep shaft process attractive.[39]

Figure 5 shows improved sludge processing for an activated sludge plant. Figure 6 shows a two-stage activated sludge process.

Aerated Lagoons

The technique of aerated lagoons was developed from adding artificial aeration to existing waste stabilization ponds. The aerated lagoon process is essentially the same as the conventional extended aeration

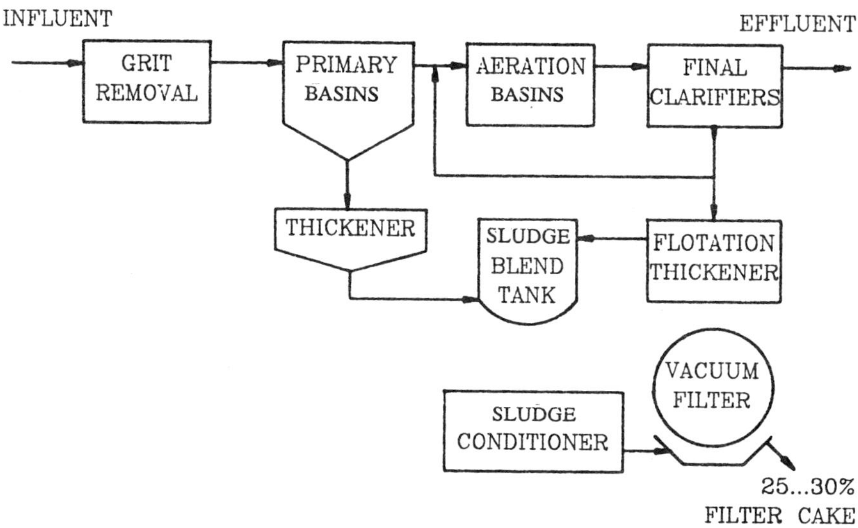

Figure 5. Improved sludge processing for an activated sludge plant.

process except that an earthen basin (for treatment of industrial wastes, it may be necessary to line the basin) is used as the reactor and the oxygen is supplied by surface or diffused aerators. However, because lagoons are not generally as well mixed as activated sludge, a low level of suspended solids is maintained in the mixed liquor. MLVSS of lagoons range from 50 to 150 mg/l.

When aeration is not sufficient to maintain aerobic conditions throughout, a portion of the biomass on the bottom may undergo anaerobic microbial decomposition. Such lagoons are called aerobic - anaerobic of facultative lagoons.

The aerated lagoon process has been successfully operated on petrochemical wastes, textile wastes, pulp and paper mills, and refinery wastes. The process can operate on the same categories and organic species as activated sludge.[3,4]

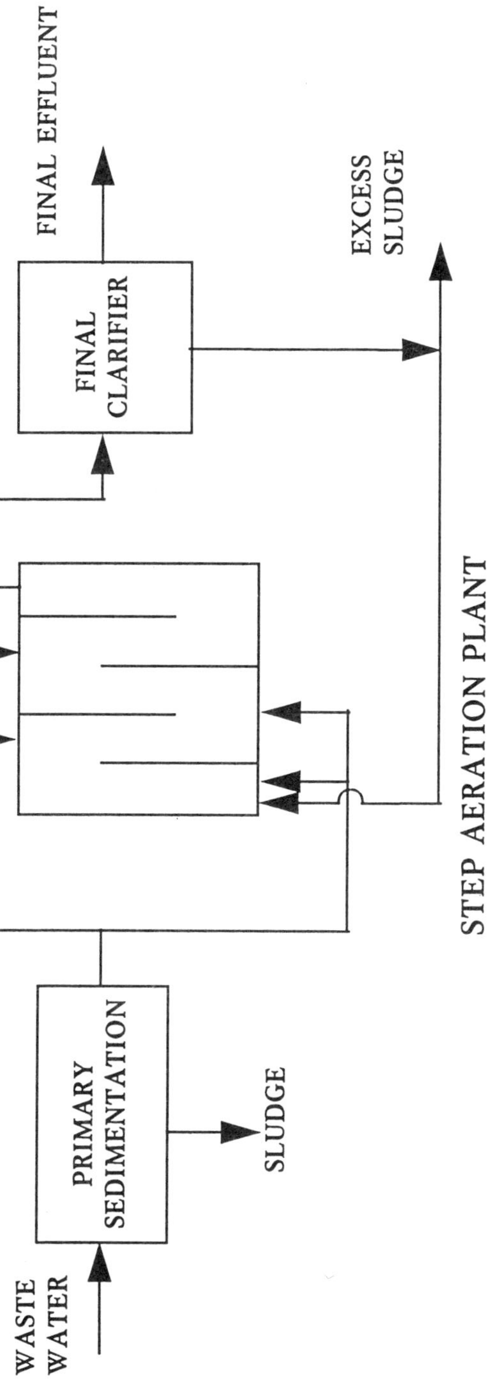

Figure 6. Two-stage activated sludge.

Waste Stabilization

Stabilization ponds are large earthen basins that treat wastes by natural processes of bacteria and algae. Aerobic conditions prevail by oxygen produced by the algae and atmospheric diffusion. To maintain aerobic conditions throughout, the contents are mixed periodically by pumps or surface aerators.

Natural biodegradation reactions are allowed to proceed as wastewater passes slowly through large shallow basins--allowing oxygenation by wind aeration (diffusion) and algae photosynthesis using sunlight energy. Waste stabilization of industrial wastes is recommended only where the waste has received preliminary treatment to remove most of the organics and essentially all the inorganics.

Waste stabilization provides aerobic and, if deep enough, facultative anaerobic decomposition of organics at the benthic sediment-water interface. Stabilization varies with changes in temperature. During cold seasons, facultative bacteria may shift to an anaerobic mode. Only simple organics, such as carbohydrates and proteins, are decomposed by anaerobic activity and the rate is less than aerobic conditions. Also, anaerobic activity increases with pond depth.

Industries employing stabilization ponds are meat and poultry, packing-cannery, dairy plants, iron and steel, oil refineries and petrochemicals plants.[3,4]

Trickling Filter (Attached Growth)

Wastes are sprayed through the air to absorb oxygen and allowed to trickle through a bed of rock or synthetic media coated with a slime of microbial growth, which is able to decompose organic matter in the waste stream.

The trickling filter relies on a media support of immobile microorganisms which receive substrate as waste and is trickled over their cell surface. The microbial community consists primarily of aerobic, anaerobic, and facultative bacteria, fungi, algae, and protozoa--some higher animals. The microbial slime remains aerobic primarily at its surface where air and water interface with cells. The underlining portion, adjacent to the media, may become anaerobic. Periodically, the microbial slime sloughs off the media and is collected and clarified from the underflow. The ability of trickling filters to accept variable hydraulic and organic loads is based on the short residence time of wastewater in the process. However, because of the short residence time, removal of organics is not as good as in activated sludge treatment.

Trickling filter treatment of aqueous wastes is proven technology for industrial waste treatment. It is especially applicable in sequence with activated sludge; the filters are more accommodating to load variations and the activated sludge unit can achieve higher organic loading efficiencies. Trickling filters are reported to successfully handle acetaldehyde, acetic acid, acetone, formaldehyde, etc. They have also been used to decompose oil and phenol.[4,41]

Rotating Biological Contactors (RBC)

Rotating biological contactors are closely packed circular disks submerged in wastewater and rotated slowly. Biological growth is attached to the surface of the disk and forms a slime layer. The disks contact wastewater and air for oxidation as it rotates. The rotation helps to slough off excess solids. About one third of the disk is submerged. The disk system can be staged in series to obtain nearly any detention time or degree of removal required. Since the systems are staged, the culture of the later stages can be acclimated to the slowly degraded materials. Only recently has laboratory disk data been effectively scaled up to predict plant operations.[36]

Packed Beds

Microbial growth can also be attached to media in a packed bed. Wastewater flow is introduced to the bottom and air or oxygen is also introduced in the wastewater.[37]

The Oxitron systems,[34] a fluidized bed wastewater treatment process, combines activated sludge and trickling filter methods. Higher volumetric loading rates are possible. Biological organisms are fixed in the system - as in a trickling filter, and are capable of providing greater process stability in handling shock and toxic loads. But there is minimal shedding of biological growth; therefore, no clarifier or recycle of sludge is required.

The key to its effective treatment is the high concentration of active biological organisms in the fluidized reactor - 12,000 to 14,000 mg/l MLVSS as compared to 1500 to 3000 for conventional activated sludge and 3000 to 6000 mg/l for pure oxygen activated sludge treatment. One pilot plant in Iowa treated influent of 3000 mg/l BOD from a corn wet milling plant. Results showed 95% BOD removal.[34]

Landfarming

Landfarming can be defined as controlled application and cultivation of wastes on soil at a properly engineered site in order to use microorganisms naturally present in the soil to decompose the organic fraction of the wastes. The process uses minimum energy, is relatively odorless, can be repeated at frequent intervals, and is not an eyesore, since it resembles a plowed field.

Landfarming is applied to the top layers of soil where natural bacteria are found in abundance. An underlying layer of impermeable clay-like soil must be in place to prevent vertical movement of the waste prior to destruction.[45] Landfilling, in turn, is not as suitable for organic waste if covered by more than a few feet. The organics will not degrade rapidly in the absence of those bacteria found in abundance only in the upper layers of soil.

This system is used to dispose of meat packing wastes, cannery waste, organic chemical wastes, and oily wastes of petroleum refineries. Waste materials with high metal content or other non-biodegradable materials are not suitable for this process.[45]

Monitoring of the site is necessary to ensure both maximum efficiency of biodegradation and compliance with environmental regulations. Samples of the soil are analyzed periodically and some sites require monitoring wells to analyze the groundwater.

Landfarming oily sludges and many organic industrial wastes may deposit heavy metals in the soil. Therefore, the land may be rendered unsuitable for growing crops for human or animal consumption. However, the increased humus content makes the site suited for growing trees and grass.

Oily wastes have been disposed of since 1959. Exxon had an eight acre landfarm at its refinery in Bayway, New Jersey. Runoff is controlled by contouring.[32,35] Landfill gas production during the biological degradation process is illustrated in Figure 7.

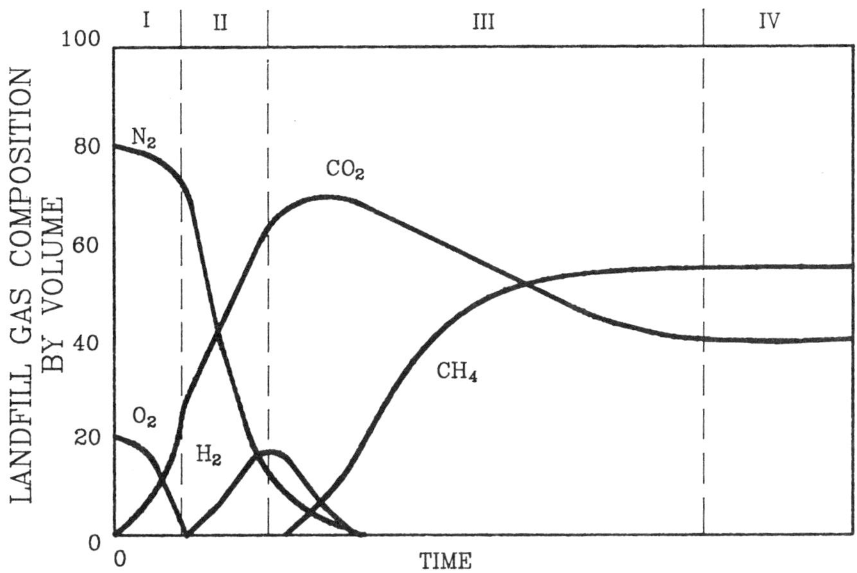

Figure 7. Landfill gas production during the biological degradation process.

Anaerobic Digestion (Treatment)

Anaerobic digestion is a process for degradation of organics in an air-free environment. Anaerobic organisms utilize part of the substrate for cell growth and the other part to produce methane and CO_2 gas. Anaerobic digestion is applicable primarily to simpler organics--carbohydrates, lipids, proteins, alcohols, and organic acids. Little degradation of long chain or cyclic hydrocarbons takes place.

The process has been mostly used for sewage sludge digestion, but should be considered for treatment of industrial wastes--such as meat packing wastes and brewery wastes.

The process relies on two types of organisms in a symbiotic relationship--acid forming and methane forming bacteria. The process produces methane and results in a low production of sludge. Typically, the process treats 5 to 7% solids; however, a modification, the anaerobic contact process is used for wastes with low solids concentration. Other modifications are the upflow contact process and the anaerobic filter process.[21,38]

Anaerobic digestion is affected by influent waste composition, pH, and temperature. Overall rate, however, is controlled by the conversion of acid, methane, and CO_2. Therefore, the methane bacteria are rate limiting microorganisms in the population.

The inhibitory effect of certain hydrocarbons on unacclimated digestion was studied by Union Carbide. Acclimation of the process to the organic inhibitor was slow and most unsuccessful.[4]

In industry, anaerobic digestion has been employed by the meat packing industry. Potentially, treatment can be applied to feedlot wastes, wastes from pharmaceutical industries, and food processing wastes.[4]

A new commercially successful anaerobic wastewater treatment process has been introduced by Celanese, called Celrobic. It is targeted for a large number of industries including chemical, petrochemical, food, brewery, and distilleries. Celanese claims that rising energy prices have made aerated processes such as activated sludge and aerated lagoons expensive. The basic unit of Celrobic is the reactor, a packed bed through which water circulates upward. The organic material is absorbed and degraded by the bacteria. To avoid microbe wipeouts that have plagued previous anaerobic systems, a computer system constantly monitors environmental factors, such as temperature and acidity-alkalinity, and makes corrections when factors get out of balance.

Celanese has not yet come up with a method to keep a reserve of trained bugs just in case wipeout occurs despite the computer. It takes two to three months to acclimate microorganisms to high concentrations of organic chemicals that generally threaten anaerobic microbes. They are, however, trying to use a freeze dried method now used for aerobic organisms so that the organisms can be reconstituted when needed.[28,29]

In a process, "Bioenergy," marketed by Capital Plant International (London), streams having a high waste content (4-10% biodegradable matter) can be converted to methane gas with much lower energy input. The major difference in the system is a patented digester that mixes incoming wastes with a sparging stirrer. The process was developed by Biomechanics, Ltd. and has been tested at commercial scale throughputs in three European locations. Throughput can be as high as 100 tons/d of COD.[15]

MUTANT BACTERIA

Mutant bacteria additives are now available to augment the naturally developed bacterial population of activated sludge, trickling filter, or lagoon treatment plants. A regular schedule of bacterial additions can provide faster system response to such problems as start-ups, plant upsets to variable and shock loads, and cold weather operations. Use of the specialized bugs seems to be increasing; later on in this chapter, several laboratory and plant applications called from the literature are noted. Bacterial cultures are sold either as liquid preparation, or as an air dried or freeze dried product. The formula may contain bacteria specific to a particular waste material, or a number of bacteria for treating a general class of waste.[12]

General Environmental Science Corporation offers a patented liquid product called LLMO (for liquid live microorganisms). The live bacteria come in a dormant state suspended in sodium sulfite which acts as an inhibitor.[6] There are seven different strains, including aerobic and facultative bacteria, for treating most organic wastes. The strains can also be used to prevent foul odors from being produced by competing with those bacteria that *can* produce odorous fumes. Shelf life is guaranteed for two years; reactivation is by dilution. Polybac Corporation selects bacteria on their ability to survive environments containing uncommon organic compounds. The bacteria are exposed to increasing concentrations of these compounds to increase tolerance of subsequent strains. Use of radiation helps to obtain strains with even more enhanced properties, namely, bacteria with specialized waste disposal capabilities. The Sybron/Biochemical Company has developed a new series of bacterial cultures, the Bi-Chem 1000 series, to aid in the biological oxidation of industrial wastewater. This series of specialized strains of microorganisms, both liquid and solid, are selected cultures designed to degrade chemical wastes and metabolize these wastes at accelerated rates under conditions which may exist in industrial waste streams. These cultures break down chemical wastes which are considered to be non-biodegradable with existing sewage bacteria.

Since the ability of the cultures to survive is marginal when introduced in a wastewater treatment plant, companies suggest a regular reinoculation schedule. This means the bacteria will be present even if their substrates are in low concentration or absent. The bacterial mixes also need a proper nutritional balance. General Environmental Science, for example, insists that wastewater contain at least 5 ppm nitrogen (as ammonia) and 1 ppm orthophosphate for every 100 ppm BOD.

Case Histories

1. At Exxon's 1 mgd activated sludge plant at Benicia, California's oil refinery, control tests of bacterial addition have been conducted. At the plant, which has two parallel trains, specialized bacteria were added (in freeze-dried form) to the east aeration basin--while the other basin remained conventional. It was found that in normal operation there was a 32% improvement in the performance of the activated sludge system on the basis of organic matter removed. Phenol and ammonia effluent levels were lower than the controlled unit. Upon seeding the west basin, Exxon obtained similar improvement. Other benefits observed were faster unit start-ups, more stable operation with variable loads, and less foaming. The saving in antifoaming chemicals exceeded the cost of the bacterial additives. Exxon concluded that regular maintenance dosages of bacteria are not necessary for oil refineries with constant crude processing.[12]

2. J.T. Baker Company's three million gallon per day secondary treatment plant at Phillipsburg, New Jersey started using bacterial cultures in 1978. Two dried cultures were used; one for ammonia removal and the other for hydrocarbon degradation. Daily dosages of 2.5 pounds and 5 pounds were used, respectively. The company believes that the cultures have improved removal efficiency. BOD levels which were close to 18 ppm were reduced to 6 ppm, and ammonia levels, which had been in the 40 to 45 ppm range, were well below 20 ppm.[12]

3. On March 25, 1982, a railroad tank car spilled 20,000 gallons of formaldehyde and contaminated the ground near Ukiah, California. Mutant bacteria had reduced the residual formaldehyde in the final stage of soil cleanup from 1000 ppm to 50 ppm after 15 days of treatment. Polybac, who made the microorganisms used in the

cleanup, also used bacteria for residual cleanup of soil contaminated with pentachlorophenol. And Polybac claims it will soon demonstrate the ability of mutant bacteria to detoxify PCBs in a test sanctioned by EPA. They claim also that bacteria can be used on other aliphatic and aromatic hydrocarbons, halogenated aliphatics and aromatics, cyanides, and nitriles.[20,23]

4. Application of mutant adapted bacteria to potentially toxic wastes prevented the release of high levels of phenol to the Ohio River. A wipeout of 4.5 mgd biological waste treatment plant caused the backup of untreated waste containing up to 60 ppm of phenol. An equalization basin and 6M gallon emergency impoundment lagoon were almost filled with contaminated wastewater. The use of mutant bacteria prevented the plant from being out of compliance with EPA effluent standards. Bacteria used was Phenobac, which is marketed by the Polybac Corporation, and is a mixture of bacterial strains.[26]

5. In 1980, about 20,000 gallons of dioxin leaked from a tank car in Sturgeon, Missouri causing the evacuation of 650 residents. The hazardous material, plus contaminated runoff, were collected in an impermeable lagoon by a spill control contractor, O.H. Material of Findlay, Ohio. The lagoon was turned into a temporary waste treatment system where nutrients and aeration equipment were added and inoculated with mutant adapted bacteria. This formulation was also Phenobac. The orthochlorophenol content of the lagoon was initially greater than 600 ppm; after one month of treatment with Phenobac, the dioxin level was down to 25 ppm.[24]

6. At the Texas Oil and Chemical Terminal in Vidor, Texas, about 13M gallons of wastewater, high in organic matter, accumulated without discharge. A means of treating this water was urgently needed; it had reached a high level in a lagoon and did not meet discharge criteria. In October 1980, aerators were added, and in January 1981, a section was seeded with a bacterial culture at the rear end of the lagoon and circulated to the front section. Within three weeks of the seeding with bacteria, the COD dropped to about 300 ppm even with low temperature prevailing. Within five months of starting the aerators and three months after seeding with bacteria,

the lagoon water was suitable for discharge. The bacterial culture, which was supplied in dried form, was very effective in consuming organic matter, and once the biological community was established, no daily maintenance additions were needed.[25]

7. Special bacteria have also been applied to a sewer system to remove grease. They were first injected in high doses to give the dominance that they must have over naturally occurring strains. A specially formulated bacterial culture was used for this purpose, DBC plus, developed and supplied by the Environmental Cultures Division of Flow Laboratories. This bacterial approach has also been used to reduce BOD in sewage treatment plant effluents and to reduce hydrogen sulfide odor problems in fly ash holding ponds.[5]

8. Mutant bacteria have also been used to control filamentous growth in mill wastewater treatment. A treatment system was developed which, in a two month trial, increased a waste treatment plant's BOD removal rates, improved sludge settling, and increased tanning and lining removal. The system, after initial acclimatization, was able to meet state mandated permit parameters under normal operating conditions. Problems were experienced with the build-up of filamentous organisms during periods of high temperature, or when the normal biomass was in weakened condition. Subsequently, the biomass was able to be acclimated so as to resist the changes that had previously caused such drastic effects.[7]

9. A two acre lagoon with 2M gallons of waste oil was located behind a bearing plant in southern New England. Floating in the lagoon was an oil layer several feet deep. Such large scale waste oil accumulation plagues many metal working plants, and there is increasing pressure to find efficient disposal methods that comply with regulations. Many of these methods are expensive. But the bearing plant found a relatively inexpensive solution. By using mutant bacteria, the contamination in the oily wastewater was reduced to levels acceptable for discharge to the municipal treatment system.[8]

10. A rendering plant with a small but concentrated organic nitrogen loading achieved a 99.9% removal of ammonia after inoculation

with a bacteria produced by in-situ genetic engineering methods. Influent BOD of over 1200 mg/l with influent ammonia over 1100 mg/l were both reduced to less than 5 mg/l in effluent quality during a program based on the use of two mutant cultures.[8]

11. An ethoxylated non-phenol detergent, which is also considered an oil and grease, was the causative agent in a foaming problem at a Taxtex, South Carolina textile plant. In a series of experiments, several organisms were isolated from samples of wastewater taken from the plant's aerated lagoons. In a batch biotower experiment using the three most effective cultures, the initial detergent concentration of 520 mg/l was reduced by 88% after 41½ hrs. Onsite treatment with the most effective of the three cultures produced a temporary but fairly rapid reduction of foam. Installation of a baffle in the tower may correct the problem and introducing the culture in a lagoon, where it could have more time to act, may provide more positive results.[9]

12. North Chemical Company of Mariett, Georgia emits process wastewater effluents to the company's treatment facility. The effluents include emulsified petroleum waxes and polyacrylic and polyvinyl acetate polymers. The waste treatment facility contains a bio-oxidation tower which is 32 ft. high and contains a honeycomb PVC media providing 3.6 acres of surface area. Its capacity is 30,000 gallons per day while the rest of the system is 50,000 gpd. Phenobac, commercial bacteria, was seeded into the biotower. Five pounds of the Phenobac were reconstituted with warm water and added to the final pH adjustment tank at the same point as the nutrients. Additions of one pound per day were made over the next ten days and 1/5 lb. was added daily thereafter for population maintenance. Within one month COD was reduced to 57% of the former average. Within three months the COD was reduced 97%.[10]

13. A study was made to investigate the feasibility of bioengineering a dairy waste treatment system and improve its performance through inoculations of selected bacteria cultures. Parameters were monitored to follow system removal efficiency, sludge floc formation characteristics, microbial population shifts; and other

functional performance parameters of the system (mainly settling and quality of final effluent). Prior to the bacterial inoculation program, the poorly operating dairy extended aeration system had a BOD removal efficiency ranging from 80 to 90% with daily BOD loadings of 900 to 1100 lbs. After the inoculated bacteria established dominance, BOD removal efficiency increased to about 98%. BOD removal efficiency was improved from a range of 70 to 80% before inoculation to about 91%. Sludge volume indexes were reduced from average values of 350 to values of 200 ml/g which reflect the improved sludge settleability and biomass compaction. Effluent discharge quality was improved by a reduction in suspended solids from averages of 200 mg/l prior to inoculation to consistent values of less than 90. Included with the improved suspended solids discharge reduction was increased solids removal efficiency of the in-line centrifuges.[13]

14. Some research organizations are working to create new microorganisms with voracious appetites for some of the most highly noxious pollutants. Battelle Memorial Institute (BMI) is focusing on bacterial means of destroying such notoriously resistant chemicals as DDT and 2,4-D. The University of Illinois is developing microbes that digest PCBs and 2,4,5-T. SRI International is trying to engineer a microbe to digest certain chemical intermediates, dye stuffs, and parathion. BMI is in the process of transferring the gene for 2,4-D degradation from a less active organism into a bacterium of the genus Pseudomonas. This organism converts 2,4-D into harmless products such as carbon dioxide and water. SRI had accumulated cultures of naturally occurring bacteria that degrade highly resistant chemicals. These strains and recombinant-DNA techniques will be exploited to obtain more active microbes for commercial use. The University of Osaka has an extensive collection of wild bacterial strains with the ability to degrade organic mercury compounds.[14]

15. The concept of bacterial augmentation was evaluated in the Northeast Sewage Treatment Plant in St. Petersburg, Florida. Treatment began in January of 1977 by adding 1 ppm of a mixed bacterial seed culture to three lift stations. By October 1977,

effluent BOD and SS levels were reduced from 1500 lb. to 391 lb. and 502 lb., respectively. Nitrite and nitrate levels were reduced to less than 0.1 mg/l and grease accumulation and odors were eliminated. When bacterial augmentation was discontinued, effluent concentration of BOD and suspended solids increased insignificantly within 60 days. The actual effect of bacterial augmentation on the plant, which bacteria was most effective, and why the improved removal deteriorated after bacterial augmentation stopped, remained to be determined.[6]

16. V. Boroneko investigated the rapid destruction of mixtures of organic substances present in industrial wastewater mixed with cultivated bacteria with an unlimited flow of oxygen and substrate in a laboratory flow-through fermenter with a mechanical aerating device. The dosage of solution, pH, and temperature were regulated automatically. Ammonia and potassium phosphate were added at a ratio of COD:N:P of 100:5:1. Traditional methods of biological purification indicated the acceptable load on a single aerated volume can be increased by greater than or equal to one order, compared with those generally accepted for traditional systems. Highly active populations of microorganisms were selected that were capable of quickly adapting to the destructive compounds. The significant exothermic effect in the destruction of pollutants created the high temperatures (35 to 37°C) necessary without additional heat.[22]

17. Batelle Pacific Northwest Laboratories is seeking to develop a genetically engineered microbe that can produce useful materials from chemical plant waste. They expect to develop the microbe in 1983 and a process for using it two years later. They are also examining use of microorganisms to accumulate metals, which might be used in cleanup of metal laden industrial wastes.[16]

18. The "bioaccumulation" of wastes is also being investigated under a joint venture of Polybac Corporation and the O'Kelly Companies. Conceptually, mutated organisms that selectively accumulate metals such as cadmium, lead, or molybdenum would be mixed into a waste stream and later separated out leaving a purified material.[17]

19. Auckland University and Warkato University (New Zealand) are exploring applications of high temperature resistant bacteria thought to exist only in geothermal pools of North Island. The bacteria, which thrive between 82 and 104°C, produce protease, lactase, and cellulose enzymes that may be useful in waste treatment and other applications.

Dissenting Opinions

The claims of manufacturers of mutant bacteria are not easily verified and experts are divided in their opinions on the cultures' effectiveness. For example, Richard Raymond, an environmental consultant with Suntech Incorporated says that "most systems (conventional) are well inoculated; they just need a good nutritional balance for each waste being degraded. There is little evidence, published and documented, that special cultures are needed."[12] While basically agreeing with Raymond, Ed Barth, chief of biological treatment at the U.S. Environmental Protection Agency's Cincinnati Research Laboratory, conceded "there may be a place for special bacterial additions in *specialized* systems." More positively, Ed Stigall, Director of Office of Quality Review, EPA's Effluent Guidelines Division, stated: "Bacterial additives will speed up the process of one culture dominating the system, and they aid plant recovery after an upset. As such, they are a very useful operations and maintenance tool for help on infrequent bad days."[12]

R.E. Grubbs of Flow Laboratories, Inc. states that "unfortunately many hazardous waste spills involve mixtures of all kinds of chemicals. With pure chemicals, like a rail car spill, it is relatively easy to develop a bug for the job." "The complexity of the chemicals and the surrounding environment can pose a threat to successful use of bacteria in the cleanup of abandoned hazardous waste sites," says W. Studebaker of O.H. Materials Company.[23]

W. Sun of Durham, North Carolina is also skeptical about the efficacy of bacterial additions. "They are solid in microgram and milligram quantities for applications that include huge volumes of waste. It is unwise to expect these preparations to cure the ills associated with a treatment process that requires strict control conditions for its success."[11]

Betz Laboratories, Inc. researchers tested a variety of cultured microorganisms that are commercially produced and sold to treat wastewater contaminated with organic compounds not normally found in nature. In bench scale, pilot plant, and commercial-scale activated sludge systems, they found little improvement in handling shock loads of phenol and no improvement in reduction in TSS of 30D. Higher than recommended dosages of the additives (above 60 lb/mg wastewater) resulted in noticeable improvement, but Betz questions the cost-effectiveness. "Our general conclusion is that over the long term, these biological additives may have some benefit-especially when genetic engineering techniques (to tailor organisms to specific wastes) become widespread--but right now it seems generally uneconomical to use them" says Steven Spoerle, technology manager at Betz.[18]

Considerable resistance toward biological augmentation in wastewater treatment is prevalent among consulting engineers, superintendents or operators of wastewater treatment plants, since they are trained that augmentation is not necessary.[44] It is difficult to convince an operator or consulting engineer of the effect of the cultures since it is difficult, if not impossible, to run carefully controlled experiments on wastewater treatment plants. Improvements and innovations will occur in treatment plants only with the combined efforts of those experts in design and operation and those in the biochemical disciplines. Certain over-statements have been made by biologists as to the effects of biological augmentation to wastewater treatment and by engineers as to the effect on plant efficiency. Answers will only be provided when those who are really involved meet and discuss the problem. So far what has been written only demonstrates that neither side has the answers to the questions firmly established.

REFERENCES

1. Gaudy, A., and E. Gaudy. *Microbiology for Environmental Scientists and Engineers*. New York: McGraw-Hill, 1980.

2. Sawyer, C.N., and P.L. McCarty. *Chemistry for Environmental Engineering*. New York: McGraw-Hill, 1967.

3. Metcalf & Eddy, Inc., and G. Tchobanoglous. *Wastewater Engineering: Treatment, Disposal, Reuse*. New York: McGraw-Hill, 1979.

4. De Renzo, D.J. *Unit Operations for Treatment of Hazardous Industrial Wastes*. New Jersey: Noyes Data Corporation, 1978.

5. Anonymous: "Clean That Sewer With Bugs," *E.S. & T.*, Oct. 1979, 13 (10), pp. 1198-1199.

6. F.I. Horsfall. "Bacterial Augmentation of Wastewater Treatment," *New England Water Pollution Control Association Journal*, Sept. 1979, 13 (2), pp. 158-163.

7. Zitrides, T.G. "Mutant Bacteria Control Filamentous Growth in Mill Wastewater Treatment," *Pulp & Paper*, Feb. 1980, 54 (2), pp. 172-174.

8. Anonymous: "Mutant Bacteria Clean Up Waste Oil," *American Machinist*, July 1981, 125 (7), pp. 104-105.

9. Davis, L., J. Blair, and C. Randall. "Mixed Bacterial Cultures Break 'Non-Biodegradable' Detergent," *Industrial Wastes*, May-June 1979, 25 (3), pp. 26-27.

10. Zitrides, T.G. "Mutant Bacteria Overcome Growth Inhibitors in Industrial Waste Facility," *Industrial Wastes*, Sept.-Oct. 1978, 24 (5), pp. 42-44.

11. Sun, W. "How Super Are the New Super Heroes?" *Water and Wastes Engineering*, Oct., 1978, 15 (10), pp. 25-46.

12. Deutsch, D.J. "Waste Treatment Boosted by Bacterial Additions," *Chemical Engineering*, Apr. 23, 1979, 86 (9), pp. 100-102.

13. Chambers, J.V. "Bioengineering an Activated Sludge Microflora to Improve Waste Removal Performance," *Fifth Annual Industrial Pollution Conference*, Atlanta/Georgia, Apr. 19-21, 1977.

14. Anonymous: "Building 'Superbugs' for the Big Cleanup," *Chemical Week*, July 23, 1980, 127 (4), pp. 39-40.

15. Anonymous: "An Anaerobic Digester That Converts Industrial Wastes to Methane," *Chemical Engineering*, May 31, 1982, p. 19.

16. Anonymous: "Batelle Pacific Moves Into the Genetic Engineering Field," *Chemical Engineering*, June 14, 1982, p. 19.

17. Anonymous: "Microbial 'Bioaccumulation' of Metals From Waste Streams," *Chemical Engineering*, Aug. 23, 1982, p. 18.

18. Anonymous: "Are 'Superbugs' Effective in Treating Wastewater?" *Chemical Engineering*, Apr. 19, 1982, p. 20.

19. Anonymous: "New Zealand Bets $1 Million on the Commercial Potential of Unique Bacteria," *Chemical Engineering*, May 3, 1982, p. 19.

20. Anonymous: "Mutant Bacteria Helped in Cleaning Up a California Formaldehyde Spill," *Chemical Engineering*, Aug. 9, 1982, p. 18.

21. Olthof, M., and J. Olezkiewicz. "Anaerobic Treatment of Industrial Wastewaters," *Chemical Engineering*, Nov. 15, 1982, p. 121.

22. Boroneko, V., L. Usmanova, and R. Naumova. "Rapid Biological Purification of Wastewater," *Khimicheskara Promyshlennost* (Russian), 1979, No. 8, pp. 468-469.

23. Anonymous: "Bugs Tame Hazardous Spills and Dumpsites," *Chemical Week*, May 5, 1982, p. 49.

24. Voras, N. "Mutant Bacteria Clean Up Herbicide," *Pollution Engineering*, Mar. 1980, p. 23.

25. Winchester, C.T., and M. Isaacs. "Aerators/Bacteria Team Up to Treat Wastewater in Refinery Lagoons," *Chemical Processing*, Sept. 1982, p. 24-25.

26. Anonymous: "Mutant Bacteria Prevent Release of Toxic Substances," *Pollution Engineering*, May 1979, p. 21.

27. Robertson, J., W. Cowen, and J. Longfield. "Water Pollution Control," *Chemical Engineering*, June 30, 1980, pp. 102-119.

28. Anonymous: "Anaerobic Treatment Cuts Costs, Generates Energy," *Chemical Week*, Aug. 5, 1981, pp. 40-41.

29. Anonymous: "For License, Anaerobic Wastewater Process," *Chemical and Engineering News*, Aug. 17, 1981.

30. Barber, N.R., and C.W. Dole. "Increasing Sludge-Digester Efficiency," *Chemical Engineering*, July 17, 1978, pp. 147-149.

31. Galdieri, J.V. "Biological Phosphorus Removal," *Chemical Engineering*, Dec. 31, 1979, pp. 34-35.

32. Huddleston, R.L. "Solid-Waste Disposal: Landfarming," *Chemical Engineering*, Dec. 31, 1979.

33. Thibault, G.T., and K.D. Tracy. "Controlling and Monitoring Activated-Sludge Units," *Chemical Engineering*, Sept. 11, 1978, pp. 155-160.

34. Anonymous: "Space-Saving Wastewater Treatment," *Chemical Engineering*, July 2, 1979, p. 47-48.

35. Frandsen, C.F. "Landfarming Disposes of Organic Wastes," *Pollution Engineering*, Feb. 1980, pp. 55-57.

36. Dallaire, G. "Behind the Rapid Rise of the Rotating Biological Contactor," *Civil Engineering*, Jan. 1979, 49 (1), pp. 72-73.

37. Holladay, D.W. "Biodegradation of Phenolic Waste Liquors in Stirred-Tank, Packed-Bed, and Fluidized Bed Bioreactors," *Journal Water Pollution Control Federation*, Nov. 1978, 50 (11), pp. 2573-2589.

38. Mosey, F.E. "Anaerobic Filtration: A Biological Treatment Process for Warm Industrial Effluents," *Water Pollution Control*, 1978, 77 (3), pp. 370-378.

39. Anonymous: "Deep Shaft-Process and Application," *Effluent and Waste Treatment Journal*, Jan. 1978, 18 (1), pp. 33-35.

40. Grieves C. "Powdered vs. Granular Carbon for Oil Refinery Wastewater Treatment," *JWPCF*, Mar. 1980, 52 (3), pp. 483-497.

41. Hosono, Y. "Trickling Filter Process Characteristic Evaluation," *Water Research*, 1980, 14 (6), pp. 580-590.

42. Beccari, M. "Nitrogen Removal from Wastewater," *Effluent and Water Treatment Journal*, Feb. 1979, 19 (2), pp. 65-71.

43. Collins, C.E. "The Effect of Temperature Control on Biological Wastewater Treatment Processes," *Water Resources*, 1978, 12 (8), pp. 547-554.

44. Horsfall, F.L. "Biochemical Augmentation of Wastewater Treatment," *JWPCF*, Feb. 1977.

45. "Protecting the Environment," *Supplement to Chemical Week*, 1982.

4 NITRIFICATION AND DENITRIFICATION IN THE ACTIVATED SLUDGE PROCESS

INTRODUCTION

The deleterious effects of nitrogenous compounds on aquatic environments have long been recognized. Nitrogenous compounds can cause a significant depletion of dissolved oxygen in receiving waters, exhibit toxicity towards fish, and therefore, decrease the productivity of streams and lakes, and present a public health hazard. To preserve receiving water quality, many discharge permits for wastewater treatment plants are being revised to include restrictions on the discharge of the various nitrogen compounds. In order to meet these permit limitations, the biological processes of nitrification and denitrification are commonly employed. Nitrification is the process by which ammonia is first converted to nitrite and, finally, to nitrate. Gaseous nitrogen is produced from nitrite and nitrate during denitrification.

Prior to 1930, most wastewater treatment facilities were designed to accomplish a relatively high degree of nitrification, at least during the warmer months of the year, since highly nitrified effluents were discovered to be unsusceptible to putrefaction.

During the period from 1940 to the late 1960's, the design of wastewater treatment facilities was directed towards the removal of carbonaceous materials in terms of the biochemical oxygen demand (BOD). Designing strictly for BOD removal was favored since it could be accomplished at reduced capital and operating costs as compared to designs incorporating nitrification.

Also, during this time, environmental engineers generally dismissed the effects of unnitrified effluents on receiving waters based on the following concepts:[1]

- Nitrification is caused by a specific group of microorganisms, the population of which is minimal in natural waters.
- The reaction rate for nitrogenous oxidation is small in relation to that for carbonaceous material.
- The oxidation of ammonia to nitrates simply converts dissolved oxygen (DO) to a form in which it is still available to aquatic organisms.

These concepts have been proven inaccurate by biologists who indicated that nitrates would not satisfy the oxygen requirements of fishes and many other aquatic fauna and by numerous river and stream investigations.

Courchaine,[2] in a study of the Grand River below the Lansing, Michigan municipal wastewater treatment plant, reported that in a stream where nitrification is active, the nitrogenous oxygen demand (NOD) exerts a demand on the oxygen resources of the stream which can result in a partial depletion of the DO assets if the rate of deoxygenation exceeds the rate of stream reaeration. Therefore, it is evident that the NOD load must be considered in any program of stream improvement having as one of its objectives, the maintenance of a given level of minimum stream DO. Table 1 provides a comparison of nitrification alternatives.

FORMS OF NITROGEN

Nitrogen has the ability to exist in seven oxidation states,[3] ranging from minus 3 to plus 5, and therefore, is found in many compounds. In wastewaters, nitrogen may be found in four forms: organic nitrogen, ammonia nitrogen, nitrite nitrogen, and nitrate nitrogen.

In fresh wastewater the nitrogen present is primarily combined in proteinaceous matter and urea as organic nitrogen. Decomposition by heterotrophic bacteria, known as ammonification, readily converts organic nitrogen to ammonia nitrogen.

Ammonia nitrogen may exist in aqueous solution as either ammonium ion or unionized ammonia. The relationship between the two forms is pH dependent and may be expressed in accordance with the following equation:

$$NH_3 + H_2O \rightarrow NH_4^+ + OH^-$$

At a pH greater than 7.0, the reaction is displaced to the left. Ammonium is predominant at any pH less than 7.0. Unionized or free ammonia in concentrations above 0.2 mg/l has been shown to be fatal to several species of fish.[3] In 1972, the National Academy of Sciences/ National Academy of Engineering Committee recommended that no more than 0.02 mg/l free ammonia be permitted in receiving waters to provide a margin of safety. Ammonia toxicity should not be a problem in receiving waters with pH below 8 and ammonia nitrogen concentrations less than about 1 mg/l.[3]

Nitrite nitrogen is unstable and easily oxidized to nitrate. It exists as an intermediate compound during the oxidation of ammonia nitrogen to nitrate nitrogen. If present in wastewater, the concentration is usually less than 1.0 mg/l. Some industrial wastes may contain nitrite nitrogen in significant concentrations.

Nitrate nitrogen is the most highly oxidized form of nitrogen. Its discharge into receiving waters from wastewater treatment plants will not result in any oxygen demand in terms of NOD. Nitrate is, however, an important nutrient for algae growth and, when present in excessive quantities, may be responsible for promoting eutrophication in streams and lakes. Thus, in certain cases, its discharge might have to be limited or prohibited to prevent excessive algae growth. In the case of potable water supplies, the maximum allowable concentration of nitrate is 10 mg/l since high concentrations (90-104 units mg/l) have been shown to cause methemoglobinemia in infants under four months old.

NITRIFYING BACTERIA

The principal organisms involved in nitrification processes are the genera *Nitrosomonas* and *Nitrobacter*. These organisms are considered to be autotrophs since they derive energy for growth and synthesis from the oxidation of inorganic nitrogen and carbon (CO_2) compounds, rather than from organic compounds. Both of these groups have rather specific environmental requirements in terms of pH, temperature, and dissolved oxygen and reproduce at much slower rates than heterotrophic bacteria. Various heavy metals and organic compounds have been found to

TABLE 1

COMPARISON OF NITRIFICATION ALTERNATIVES

System Type	Advantages	Disadvantages
Combined carbon oxidation - nitrification		
Suspended growth	Combined treatment of carbon and ammonia in a single stage Very low effluent ammonia possible Inventory control of mixed liquor stable due to high BOD_5/TKN ratio	No protection against toxicants Only moderate stability of operation Stability linked to operation of secondary clarifier for biomass return Large reactors required in cold weather
Attached growth	Combined treatment of carbon and ammonia in a single stage Stability not lined to secondary clarifier as organisms on media	No protection against toxicants Only moderate stability of operation Effluent ammonia normally 1-3 mg/l (except RBD) Cold weather operation impractical in most cases
Separate stage nitrification		
Suspended growth	Good protection against most toxicants Stable operation Very low effluent ammonia possible	Sludge inventory requires careful control when low BOD_5/TKN ratio Stability of operation linked to operation of secondary clarifier for biomass return Greater number of unit processes required than for combined carbon oxidation - nitrification
Attached growth	Good protection against most toxicants Stable operation Less sensitive to low temperatures Stability not linked to secondary clarifier as organisms on media	Effluent ammonia normally 1-3 mg/l Greater number of unit processes required than for combined carbon oxidation - nitrification

suppress or inhibit the growth of nitrifiers. *Nitrosomonas* can only oxidize ammonia nitrogen to nitrite nitrogen, while *Nitrobacter* is limited to the oxidation of nitrite nitrogen to nitrate nitrogen.

NITRIFICATION STOICHIOMETRY

The oxidation of NH_3-N to NO_3-N occurs in two steps as represented by the following equations:[5]

$$NH_4^+ + 1.5\ O_2 \xrightarrow{\text{Nitrosomonas}} 2H^+ + H_2O + NO_2^- \quad (1)$$

$$NO_2^+ + 0.5\ O_2 \xrightarrow{\text{Nitrobacter}} NO_3^- \quad (2)$$

The overall reaction may be represented by combining Equations (1) and (2):

$$NH_4^+ + 2O_2 \rightarrow NO_3^- + 2H^+ + H_2O \quad (3)$$

In order for this reaction to go to completion, 4.57 mg of O_2 are required per mg NH_4^+ -N.

Assimilative reactions are also occurring during nitrification according to Equation (4):[4]

$$NH_4^+ + 4CO_2 + HCO_3^- + H_2O \rightarrow C_5H_7NO_2 + 5O_2 \quad (4)$$

in which $C_5H_7NO_2$ is the empirical formula of a bacterial cell.

By combining Equations (3) and (4), the overall oxidation and assimilation reaction is:[4]

$$22NH_4^+ + 37O_2 + 4CO_2 + HCO_3^- \rightarrow C_5H_7NO_2 + 21\ NO^{3-} + 2OH_2O + 42H^+ \quad (5)$$

Equation (5) also indicates that alkalinity is destroyed during the nitrification process. Equation (6) shows that one mole of calcium bicarbonate is required to neutralize every two moles of nitric acid produced from the nitrification process:[6]

$$2H^+ + 2NO_3^- + Ca(HCO_3)_2 \rightarrow Ca\,(NO_3)_2 + 2CO_2 + 2H_2O \tag{6}$$

Alternatively, alkalinity destruction can be expressed as 7.14 mg/l of alkalinity as $CaCO_3$ destroyed per mg NH_3-N oxidized.

The theoretical alkalinity destruction is rarely observed in data obtained from laboratory, pilot plant, or full-scaled studies. In most cases, the actual alkalinity destroyed is less than the theoretical value. Scearce et al.[7] have reported that mineralization of organic nitrogen to NH^+ occurs in the activated sludge process and imparts alkalinity to the wastewater. This addition of alkalinity is the reason that the theoretical ratio is rarely observed. The theoretical value is only valid for wastes in which no NH^+-N is obtained from mineralization of organic-N; i.e., the TKN (total kjeldahl nitrogen):NH^+-N ratio is one. As the TKN:NH^+-N ratio increases a larger amount of organic -N is present and, therefore, a greater potential for imparting alkalinity to the wastewater exists. The net change in the alkalinity of a wastewater can be predicted by:[7]

$$\blacktriangle\text{Alk} = 3.57\,(\blacktriangle\ \text{filtrate organic N- synthesized N}) - 7.14\,(\blacktriangle\ NO_3^-\ \text{-N}) \tag{7}$$

NITRIFICATION PROCESS VARIABLES AND KINETICS

Environmental conditions necessary for the growth of nitrifying bacteria are much more specific than those of most heterotrophic bacteria responsible for carbon removal. In the succeeding portions of this section, the impact of temperature, pH, dissolved oxygen, solids retention time, ammonia concentration, organic concentration, and inhibitory compounds on the rates of growth and nitrification are examined.

Ammonium Oxidation

It is widely reported in the literature[3,8,9,10] that the oxidation of ammonium to nitrite is a zero-order reaction with respect to ammonium concentration for concentrations down to about 1 to 5 mg/l.

The rate limiting step in nitrification is the conversion of ammonium to nitrite by *Nitrosomonas*.[5,8,11,12] Using the kinetic equation proposed by Monod, the growth rate of *Nitrosomonas* under steady-state conditions can be described as:[5,11]

$$u_N = \frac{u^*N}{K_N + N} \qquad (8)$$

where:

u_N = growth rate of 9 *Nitrosomonas* days $^{-1}$

u^* = maximum growth rate of *Nitrosomonas*, days $^{-1}$,

N = NH_4 + -N concentration, mg/l, and

K_N = N concentration at which $u_N = 0.5u^*$, mg/l

The oxidation of NH^+ -N can then be related to the *Nitrosomonas* growth rate as follows:

$$q_N = \frac{u_n}{Y_N} + q_n^* \frac{N}{(K_N + N)} \qquad (9)$$

where:

q_N = ammonia oxidation rate, mg NH_4^+-N oxidized/mg VSS/day,

Y_N = organism yield coefficient, mg VSS grown/mg NH_4^+-N removed,

q_N^* = maximum ammonia oxidation rate, mg NH_4^+-N oxidized/mg VSS/day, and

K_N = N concentration at which $q_N = 0.5\ q_N$, mg/l.

A zero-order relationship exists between the oxidation rate and substrate concentration when the value of K_N is much less than N.[11]

Nitrite Oxidation

The oxidation of nitrite to nitrate is also considered to be a zero order reaction.[8] The maximum growth rate of *Nitrobacter* is considerably greater than the rate of *Nitrosomonas*. , This may account for the normally low NO^--N concentrations observed in nitrifying processes. Nitrite nitrogen concentrations greater than 20 mg/l are inhibitory towards *Nitrosomonas*.

Solids Retention Time (SRT)

The SRT is the principal factor which determines whether or not an activated sludge process will support nitrification. It may be defined by the following equation:

$$SRT = \frac{X_T}{X_r} \text{ or } X_T/X_r \tag{10}$$

where:

X_t = total mass of solids (VAS) in system, lbs.,

X_r = total quantity of solids (VAS) removed from system daily, including those solids purposely wasted as well as those lost in the effluent, lbs./day.

For a given SRT and a knowledge of the growth rate of a specific type of organism under given conditions, it is possible to predict whether the specific organism will be able to establish itself in the activated sludge culture or whether it will not be able to grow fast enough to prevent it from being washed out of the system. In order for an organism to maintain itself in the activated sludge culture, it must have a reciprocal growth rate of a certain level:[13]

$$\theta_C = \frac{1}{u} \tag{11}$$

where: θ_C = SRT, days.

Since nitrifying bacteria have lower rates than the majority of heterotroph in an activated sludge system, it is highly probable that the system will be operated with a SAT that is less than the reciprocal growth rate of the nitrifiers. In such a system, nitrification will not be possible. Therefore, if nitrification is desired, the SAT must be considered with respect to the nitrifiers rather than the heterotroph.

The minimum SAT values reported for which nitrification was observed ranged from 2 to 20 days.[13-17] The variations in reported SAT values are due to several factors: temperature of wastewater, pH, DO concentration, and system type. All of these factors and their effect on the SAT are discussed in subsequent sections.

It should be noted that the choice of a design and operating value of SAT fixes only the total system biomass for a given wastewater composition and allows design tradeoffs between the aeration tank volume and the mixed liquor volatile suspended solids (MLVSS) concentration. Thus, nitrification may be achieved at short hydraulic retention times provided that the aeration tank MLVSS concentration and solids wasting policy are chosen correctly to maintain the selected SAT value.[15]

Effect of Temperature on Kinetics

While nitrification occurs over a wide temperature range, reduction of temperature results in a reduction of the maximum specific growth of the nitrifying bacteria.[9] Therefore, in order to maintain high levels of nitrification at temperatures below 15°C it is necessary to increase the SAT by as much as five times.[1,9,17] If nitrification is desired throughout the entire year, the minimum expected temperature, and the resultant minimum specific growth rate constant must be considered in the design.[9,18] The effect of temperature on the growth rate of *Nitrosomonas* at any effluent ammonia nitrogen concentration can be illustrated by an equation proposed by Young et al.:[17]

$$u = 0.18\, S \exp(0.116(T-15))/(S + 10^{0.051T-1.158}) \tag{12}$$

where:

T = *temperature, °C, and*

S = *NH_3-N effluent concentration, mg/l.*

Effect of pH on Kinetics

There is a wide range in the reported pH optima (pH 6.5 to 8.6).[1,8,10,19] However, there is general agreement that as the pH shifts to the acid range, the rate of nitrification declines.[5] Thus, it is important that sufficient alkalinity be present in the wastewater to prevent a significant decline in the pH. It is recommended that a residual alkalinity of 50 mg/l for aeration and at least 150 mg/l for high purity oxygen systems be maintained for pH control during nitrification.[10,20] Low pH conditions are only inhibitory and not toxic toward nitrifiers. Caustic or lime addition may be required to supplement low alkaline wastewaters.

Benninger and Sherrard[21] reported that the quantity of alkalinity destroyed increases as the SAT increases. This was attributed to the increased nitrification rates observed at higher SAT values.

The relationship between alkalinity destruction and the COD:TKN ratio has been reported by Scearce et al.[7] This study indicated that at high COD:TKN ratios little destruction of alkalinity can be expected, since a large portion of TKN (total kjeldahl nitrogen) is channeled into the microbial mass through synthesis reactions and is unavailable for nitrification reactions. Conversely, at low COD:TKN ratios, a smaller portion of nitrogen is incorporated into microbial cell structure and, therefore, a larger amount is available for nitrification. Hence, greater alkalinity destruction and a larger pH drop can be expected at low COD:TKN ratios.

Effect of DO on Kinetics

Because of the high oxygen requirements for the conversion of ammonia to nitrate (4.57 mg O_2/mg NH^+-N oxidized), an actively nitrifying activated sludge process requires a larger quantity of process air to sustain the same DO as a non-nitrifying process, all other factors being equal. Nitrifying bacteria are more sensitive to DO conditions than the majority of heterotroph found in activated sludge. Although there is disagreement in the literature as to the minimum DO concentration required for nitrification to proceed, it is generally agreed that 1.5 to 2.0 mg/l of DO are necessary for optima nitrification rates. The Monod Equation may be applied to determine the relationship between a given DO concentration and its effect on the maximum specific growth rate of the nitrifying bacteria.[5] Nitrification will proceed at dissolved oxygen concentrations as low as 0.2 mg/l.[22] However, the rate at which it proceeds will be significantly lower than those observed at higher concentrations of DO. Therefore, at low DO concentrations, the aeration tank detention time must be increased to permit complete nitrification. Charley et al.[12] observed that high dissolved oxygen concentrations (38 mg/l) are not inhibitory towards nitrifiers following a brief acclimation period.

Effect of Organic Loading on Kinetics

The organic loading to the aeration tanks affects the efficiency of nitrification as the organic matter controls the growth of heterotroph.[19] High organic loads accelerate the heterotrophic growth rate and the sludge quantities produced in the system. Thus, any increase in the BOD_5:TKN ratio results in a decrease of the fraction of nitrifiers present in an activated sludge system. For BOD_5:TKN ratios between one and three the fraction is less than 20 percent and greater than 8 percent.[5]

Lawrence and Brown[15] reported that at a COD:N ratio of 20:1, the nitrogen present would be just sufficient to satisfy the nitrogen synthesis requirements of heterotroph. Thus, ammonia oxidation would not occur. However, at a COD:N ratio of zero, all of the ammonia present would be oxidized. Within these limits, the fraction of NH_3-N oxidized versus the fraction assimilated by heterotrophic growth would be a function of the growth rate of the system, since the amount of excess biomass produced is a function of the mean net specific microbial growth rate.

Wild et al.[1] observed no change in the nitrification rate of a laboratory scale reactor when subjected to instantaneous increases or decreases on BOD concentrations from 50 to 5 mg/l or 50 to 100 mg/l. The author did state, however, that a change in the average BOD concentration of the feed would affect that percentage of MLVSS that is composed of nitrifiers and, consequently, the rate of nitrification.

Figure 1 illustrates pretreatment alternatives for separate stage nitrification.

Inhibition of Nitrification

Hockenbury and Grady[23] reported that aniline, ethylenediamine, hexamethylenediamine, and monoethanolamine were industrially significant organic chemicals capable of inhibiting ammonia oxidation by *Nitrosomonas*. In a subsequent study by Joel and Grady,[24] it was determined that at a solids retention time of seven days significant degradation of aniline occurs, thus removing its inhibitory effect toward nitrification.

Laboratory investigations performed by Beckman et al.[19] indicated that slug doses of copper and chromium at concentrations of 1.0 mg/l or less did not retard nitrification. Slug doses of zinc and nickel at concentrations less than 0.5 mg/l reduced the rate of nitrification. Complete inhibition of nitrification was not observed with zinc and nickel concentrations as high as 3.0 mg/l. Effects of other heavy metals were not reported in the literature examined.

Hockenbury et al.[25] concluded from laboratory studies on domestic and industrial wastes that heterotrophic bacteria do not exert an antagonistic effect on nitrifying organisms, but could in fact stimulate nitrification rates by the release of their metabolic products into the surrounding medium. He also reported that, in the absence of specific inhibitors, the failure for nitrification to proceed at the head-end of an activated sludge basin is due to the operational characteristics or environmental conditions (low pH and/or DO) of the system, rather than to the presence of organic compounds or active heterotroph in the system.

Shock organic loads to nitrifying sludges have been known to inhibit nitrification. Reinhart[26] observed an increase in MLVSS concentration and reduced nitrification rates following a peak in the BOD_5 influent concentration. Severe oil and grease loadings were causing sudden increases in the organic load to the pilot plant resulting in nitrification inhibition. Stover et al.[27] reported a similar response to sudden increases in the influent COD. This study concluded that as a result of the abrupt COD increase and subsequent accelerating heterotrophic growth rates, an intermediate by-product is excreted by heterotroph having an inhibitory effect on nitrifiers.

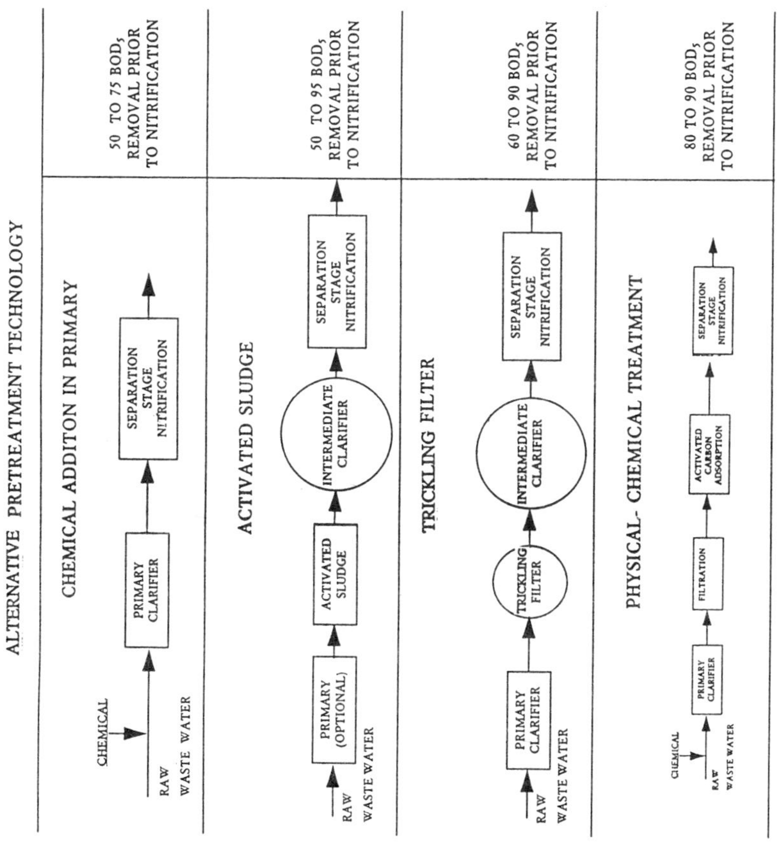

Figure 1. Pretreatment alternatives for separate stage nitrification.

Unionized ammonia (free ammonia, NH_3, FA) and unionized nitrous acid (free nitrous acid, HNO_2, FNA) can also inhibit nitrification. Studies by Anthonisen et al.[28] indicated that FA inhibited *Nitrobacter* at concentrations substantially lower than those that inhibited *Nitrosomonas* (0.1 - 1.0 mg/l for *Nitrobacter*). The ranges of FNA concentrations that begin to inhibit nitrifying organisms are between 0.22 - 2.8 mg/l. When nitrification is incomplete or will not begin, the possibility of inhibition should be investigated by determining the FA and FNA concentrations and comparing them to the above ranges. Inhibition by FA or FNA is not permanent and may be removed by adjusting operational conditions. Dilution, pH adjustment, and denitrification will reduce inhibitory concentrations of FA and FNA.

Neufeld et al.[29] examined the effect of phenol on nitrifying cultures in a laboratory fill and draw system. Concentrations of 0 - 75 mg/l of phenol were found to cause progressive inhibition of ammonia oxidation by *Nitrosomonas*.

DENITRIFICATION

Denitrification is a process by which certain species of bacteria under anoxic conditions reduce nitrate nitrogen to the gaseous end-products of N_2, NO, or N_2O which can then escape from solution to the atmosphere. Unlike other nitrogen compounds, the gaseous forms of nitrogen are relatively unavailable for biological growth and have no significant effect on environmental quality. The presence of oxidized nitrogen and organic carbon are important requisites for denitrification to proceed.

DENITRIFYING BACTERIA

Unlike the autotrophic nitrifying bacteria responsible for nitrification, denitrifying bacteria are composed of ubiquitous, heterotrophic organisms. The most common denitrifying bacteria are *Bacillus denitrificans*, *Micrococcus denitrificans*, *Pseudomonas stutzeri*, and *Achromobacter* sp.[30] In the absence of molecular oxygen, these organisms use nitrate or nitrite as terminal electron acceptors, while oxidizing organic matter for energy. This metabolic process, termed nitrate dissimilation, results in the eventual reduction of nitrate to nitrogen gas. At DO concentrations greater than approximately 0.5 mg/l, oxygen will be more readily utilized as the final electron acceptor than nitrates by the above named organisms.[31] Thus, denitrification will be inhibited when the DO concentration exceeds 0.5 mg/l.

DENITRIFICATION STOICHIOMETRY

Denitrification is a two step process and, using methanol as the electron donor, may be represented by the following equations:[6]

$$2CH_3OH + 6NO_3^- \rightarrow 6NO_2^- + 2CO_2 + 4H_2O \qquad (13)$$

$$3CH_3OH + 6NO^{2-} + 3O_2 \rightarrow 3CO_2 + 3N_2 + 3H_2O + 6OH^- \qquad (14)$$

The overall reaction using methanol may be expressed as:

$$5CH_3OH + 6NO^3- + 6O_2 \rightarrow 5CO_2 + 3N_2 + 7H_2O + 6OH^- \qquad (15)$$

Since cell synthesis occurs simultaneously with nitrate reduction, the overall nitrate removal with methanol including cell synthesis may be represented by Equation (16):[6]

$$68NO_3 + 80CH_3OH + 98H^+ \rightarrow 30CO_2 + 24N_2 + 10C_5H_7O_2N + 174H_2O \qquad (16)$$

The equation most commonly applied to determine the methanol requirements in a biological denitrification unit where nitrate, nitrite, and dissolved oxygen are present is:

$$Cm = 2.47N_a + 1.53N_i + 0.87D \qquad (17)$$

where:

Cm = required methanol concentration, mg/l,

N_a = initial NO_3-N concentration, mg/l,

N_i = initial NO_2-N concentration, mg/l, and

D = initial DO concentration, mg/l.

The biomass production (C_b, mg/l) may be calculated similarly:

$$C_b = 0.53Na + 0.32Ni + 0.19D \quad (18)$$

From equation (15), it is apparent that hydroxide is produced during the denitrification process. The hydroxide produced reacts with carbonic acid to produce bicarbonate alkalinity. The stoichiometric quantity of alkalinity produced is 3.57 mg alkalinity as $CaCO_3$ per mg of nitrate of nitrite reduced to nitrogen gas.[5] Denitrification, therefore, partially reverses the effects of nitrification whereby 7.14 mg alkalinity is destroyed per mg NH_3-N oxidized.

DENITRIFICATION PROCESS VARIABLES AND KINETICS

Like nitrification, denitrification requires specific environmental conditions to be present before it can proceed. Factors considered in subsequent sections include nitrate nitrogen concentration, pH, temperature, and carbon concentration.

Effect of NO_3-N Concentration on Kinetics

The concentration of nitrate available will affect the maximum growth rate of the organisms responsible for denitrification. The relationship between nitrate concentration and the denitrifier growth rates can be described by the Monod expression.[5]

Effect of Temperature on Kinetics

Temperature has a significant effect on the growth rate of denitrifying bacteria. It may be estimated by using the following expression:[4]

$$P = 0.25T^2$$

where:

P = percent of denitrification growth rate at 20°C, and
T = temperature, °C.

Mulbarger[32] observed a nonlinear relationship between the denitrification rate and temperature in pilot plant studies. Denitrification will proceed at reduced rates at temperatures as low as 5°C.

Effect of pH on Kinetics

Denitrification is inhibited below pH 6.0 and above pH 8.0. Although there is some disagreement in the literature as to the optimum pH for denitrification, it is generally agreed upon that the pH should be between 6.5 and 7.5.

Effect of Carbon Concentration on Kinetics

The effect of carbon concentration on the maximum growth rate of denitrifiers may be described by a Monod type of expression. The low value (0.1 mg/l of methanol) reported for the half saturation constant for methanol (K_M) indicates that large excesses of methanol above stoichiometric requirements are not required in the effluent from a suspended growth denitrification process to achieve maximum denitrification rates.[5]

NITRIFICATION PROCESSES

It is beyond the scope of this book to discuss the various modifications of the activated sludge process capable of achieving nitrification. The reader is referred to the EPA Process Design Manual for Nitrogen Control[5] for a detailed examination of the various processes. Discussion in this book will be limited to comparisons between plug-flow versus complete mix reactors and single-stage versus two-stage nitrification.

Plug-Flow Versus Complete Mix

It is generally agreed upon in the literature that a plug-flow reactor is more effective than a completely mixed reactor in maintaining a low effluent ammonia concentration. Several reasons are given to support this view. Poduska and Andrews[9] indicate that plug-flow reactors are more efficient because the minimum overall reaction rate for substrate removal is first-order. Furthermore, because the rate of reaction may vary, the plug-flow reactor may offer an additional advantage in that the reaction order will increase through the length of the reactor as the substrate concentration decreases to less than 2 mg/l of NH^+-N. The completely mixed reactor is limited in this regard because it operates a homogeneous unit in which the reaction rate is uniform throughout the reactor. The same authors also observed that a plug-flow system was also better suited to handled pulses of toxic materials. A short toxic pulse can pass through a plug-flow system, contacting only a small fraction of the nitrifying organisms. However, in a completely mixed reactor, a toxic material will be distributed instantaneously so that it contacts almost all of the organisms. Sawyer et al.[10] suggests that because the rate of oxidation of ammonia is essentially linear (zero-order reaction), short circuiting must be prevented by ensuring plug-flow conditions are present throughout the reactor. This is best accomplished by dividing the tank into a series of compartments (minimum of three) with ports between them.

Single-Stage Versus Two-Stage Systems

The current design considerations for activated sludge systems in which both BOD and NH_3-N removal is desired, center on whether the removal of both pollutants will be carried out in a single reactor (single-stage nitrification) containing both heterotrophic and autotrophic microorganisms or in two separate reactors (two-stage nitrification) in which separate and distinct heterotrophic and autotrophic populations are maintained. In two-stage nitrification, BOD removal is accomplished in a high rate activated sludge basin prior to the nitrification reactor. An intermediate clarifier between the two reactors prevents the high rate and nitrifying sludges from being mixed together. In the subsequent section, it will be seen that the selection of either system must be done on a case by case basis and will depend upon such things as the amount of ammonia removal desired, seasonal requirements, cost, and land availability.

Mulbarger[32] concluded that the separate sludge system with isolated, optimized sludge cultures was preferred over single-stage systems for nitrification. It was observed that mixed BOD removal and nitrification resulted in a loss of soluble carbon removal efficiency, longer aeration times for nitrification, and a loss of nitrogen removal caused by sludge synthesis. Barth et al.[33] indicated that the separation of carbonaceous removal and cellular synthesis from nitrification was necessary to prevent the loss of nitrifiers by washout from sludge wasting. In a separate stage, the nitrification biomass becomes an enriched culture of nitrifiers, the population of which is limited only by the ammonia concentration.

Bench-scale studies by Stover et al.[16] demonstrated that higher SAT values are required in single-stage systems to achieve comparable nitrification rates and levels of ammonia removal as in two-stage systems. A single-stage reactor with short hydraulic retention times may be unable to maintain the necessary SAT for nitrification. The maximum permissible MLSS concentration in the reactor will be determined by the solids loading rate to the secondary clarifiers. Clarifier upsets resulting in the escape of activated sludge solids to the effluent are detrimental to nitrification since they effectively increase the sludge wasting rate, thereby decreasing the SAT.

The seasonal variation in the temperature of the wastewater is another factor which should be considered before selecting the appropriate type of system. Many engineers feel that, in order to accomplish nitrification, a separate two-stage system is a mandatory requirement in northern climates where wastewater temperatures fall below 18°C. It is believed that since the sludge in the separate nitrification reactor has a significant nitrifier population, nitrification would not be as temperature sensitive as in a sludge with a marginal population. Bench-scale studies by Lawrence and Brown[15] failed to show any significant difference between the performance of both systems when operated at temperatures of 8°C and 20°C. In both systems, ammonia breakthroughs in the effluent did occur with significant increases in the ammonia mass loading at 8°C. Young et al.[17] reported that due to the poor sludge settleability observed at wastewater temperatures less than 15°C, application of the two-stage system is favored at reduced temperatures. A higher percentage of nitrifiers can be maintained in the nitrification stage of the two-stage system than in the single-stage system and nitrification can be achieved with a lower concentration of MLSS in the nitrification reactor.

Two-stage nitrification provides protection from toxic substances and shock organic loads which inhibit nitrification.[34] Before reaching the second stage intrification reactor, most of the toxics, if biodegradable, and organic materials will be oxidized in the first reactor. Also heavy metals may precipitate or be absorbed by the sludge in the first stage, thus causing no harm to the nitrifiers. However, as Lawrence and

Brown[15] point out, toxic materials might be excluded from wastewater systems by regulation rather than by relying on a "sacrificial" biosystem. Also, if phosphorous removal is required in addition to nitrification and coagulants are used prior to aeration, heavy metals will be removed in addition to the phosphorous.

In spite of the seemingly negative attributes of single-stage nitrification, it has been demonstrated to be a viable means of accomplishing nitrification[14, 15,17,19,35] and to have certain advantages over two-stage systems. Lawrence and Brown[15] take issue with the claim that two-stage systems have more positive control capabilities. They point out that the use of the SRT concept and controlled sludge wasting make the single-sludge system as controllable as the two-stage system. In the two-stage system, due to the low MLSS concentration in the nitrification reactor, poor solids separation can occur in the clarifier. This may necessitate the wasting of first stage solids into the nitrification reactor or the use of coagulants at increased operating costs. The additional clarifier required in the two-stage system results in the increased possibility of failure. Finally, because of the higher levels of sludge production and, consequently, the increased demand on sludge disposal processes, plus the additional reactor, clarifier, and land requirements, two-stage nitrification systems will have higher capital and operating costs than comparable single-stage systems.[36] Thus, it is essential that comprehensive studies of a facility be performed before selecting one of the processes. Figure 2 ilustrates a nitrification-denitrification flow sheet utilizing low porosity fine media in columns. Figure 3 shows a process schematic for submerged high porosity media columns.

DENITRIFICATION PROCESSES

The discussion of denitrification processes will focus on the kind of carbon source applied to the denitrification reactor. Additional consideration will be given to the various combinations of nitrification and denitrification processes reported in the literature.

Denitrification Using Methanol as the Carbon Source

Denitrification processes utilizing methanol are most commonly applied following a nitrification system. Refer to Figure 4. The denitrification

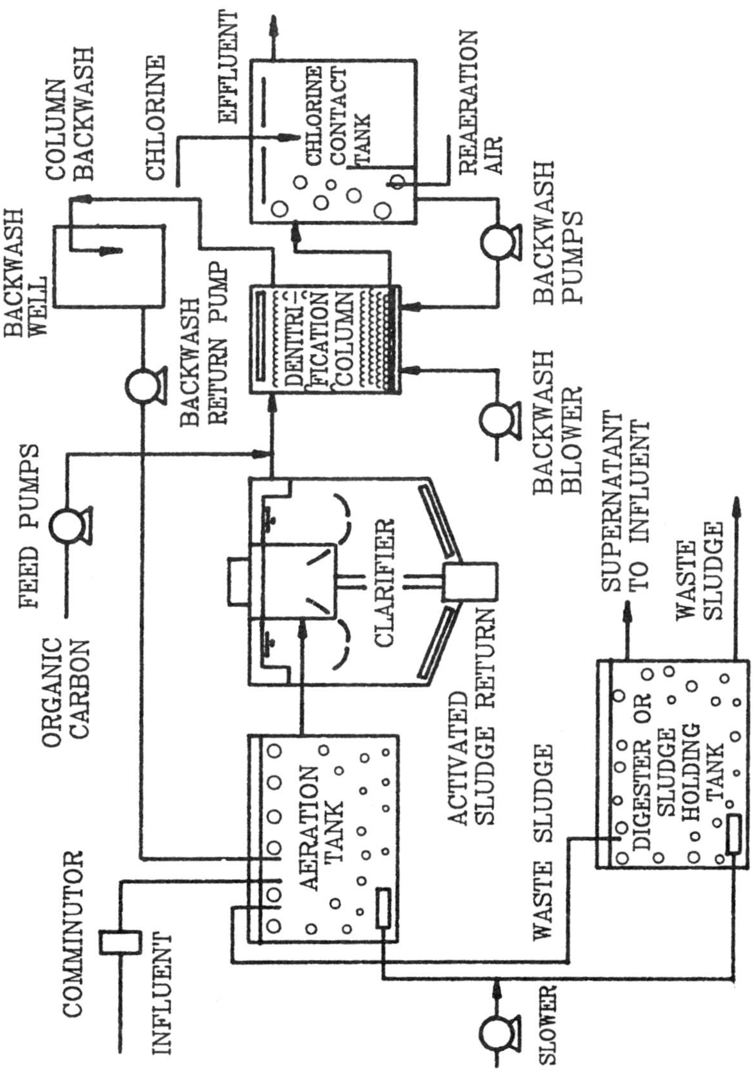

Figure 2. Nitrification-denitrification flow sheet utilizing low porosity fine media in columns.

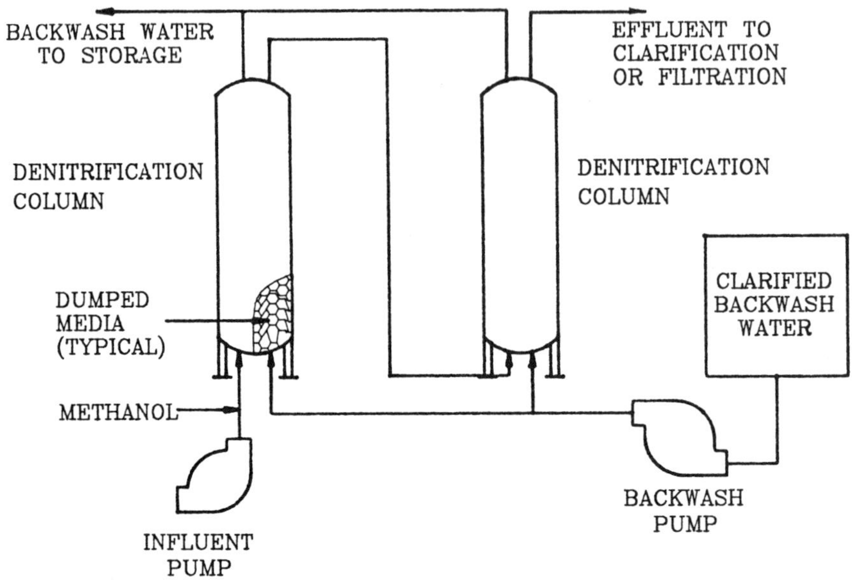

Figure 3. Typical process schematic for submerged high porosity media columns.

reactors are sealed or at least covered to prevent entrainment of atmospheric oxygen. Mixers are provided to keep the MLSS in suspension and to ensure complete dispersion of the added methanol. Following denitrification and prior clarification, the denitrification tank effluent enters an aeration basin where the entrained nitrogen gas and carbon dioxide bubbles are stripped to permit settling of the MLSS in the clarifier. The additional aeration will also remove most of the residual methanol.

Methanol is the cheapest commercial source of carbonaceous matter currently available. Glucose is the next cheapest source. Methanol is preferred over glucose because it is more completely oxidized and produces less sludge for disposal.[10] The dosage requirements for methanol are dependent on the influent NO_3-N, NO_2-N, and DO concentrations and may be computed using Equation (17) which was presented in a preceding section. It is generally agreed upon that

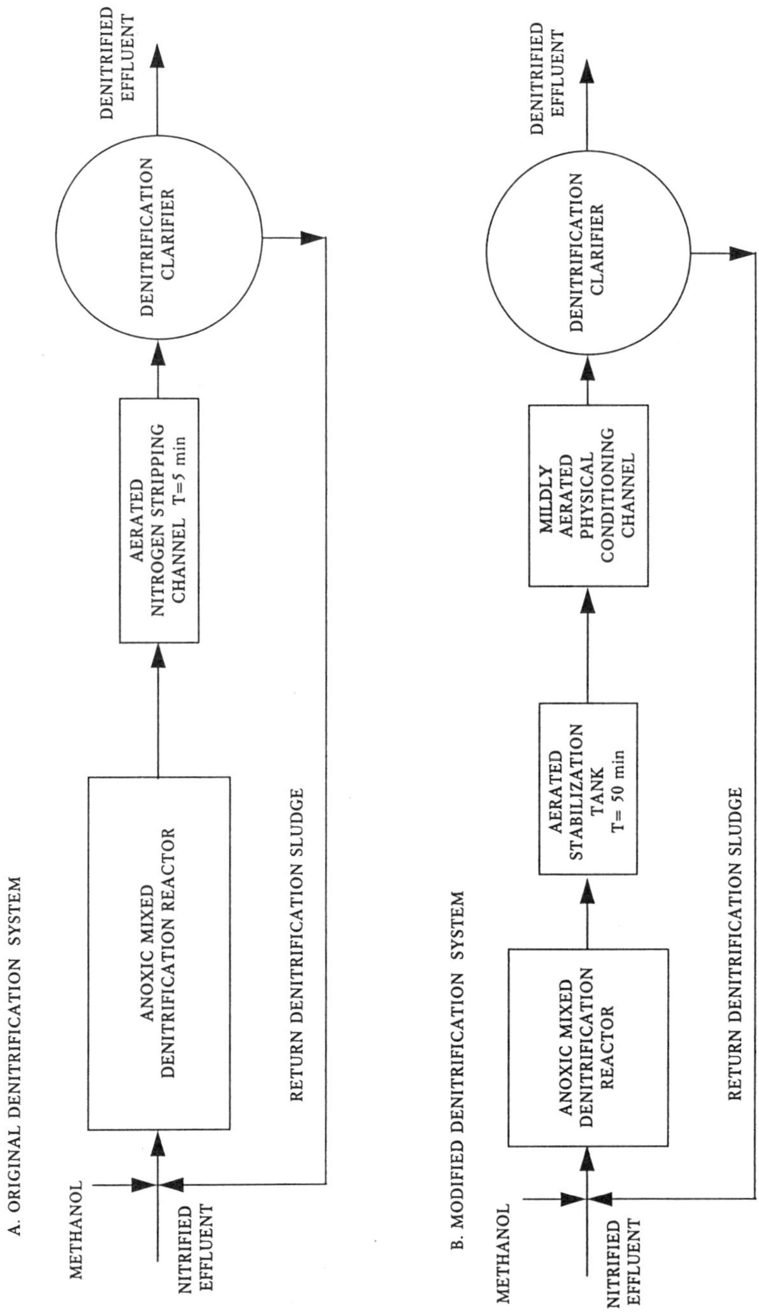

Figure 4. Suspended growth denitrification systems using methanol.

3-4 parts parts of methanol are required per part of NO_3-N for efficient denitrification. If sufficient methanol is not added to the system, the NO_3-N is mostly converted to NO_2-N.[33] Since excess methanol will appear in the effluent as BOD, automated methanol addition as a function of the mass flow of NO_3-N will maximize the efficiency of denitrification systems. According to Sherrard et al.[37] methanol requirements decrease as the SRT is increased due to a reduction in the net sludge production. Increased methanol dosages are required at higher temperatures at a constant SRT.

Denitrification Using Organic Matter Present in Raw Wastewater

Lower denitrification rates may be expected when oxidizable organic substances in raw wastewater are substituted for methanol. More importantly, when domestic sewage is applied as the organic carbon source in the flow scheme described in the preceding section for methanol, increased NH_3-N levels are observed in the effluent. Any ammonia present in the sewage will pass through the denitrification system essentially unaltered. However, because of the operating expense associated with methanol addition, flow schemes have been devised to permit the use of organics derived from wastewater while preventing ammonia bleedthrough in the effluent. A brief discussion of these processes follows. A comparison of denitrification processes is shown in Figure 5.

In a one-sludge predenitrification and nitrification process, denitrification is accomplished in the anoxic first stage with the organics present in the raw wastewater being utilized as electron donors for nitrate reduction. The conversion of ammonia to nitrate and the oxidation of any biodegradable organics not removed in the first stage is accomplished in the aerobic second stage. In this type of system, both sludge and mixed liquor are recycled. Sludge recycling is to maintain a desired level of MLVSS concentration[38] and mixed liquor recycling is required to produce a final effluent which has a low residual nitrate concentration.[39] The low BOD concentration in the denitrification effluent prevents high organic loadings to the aerobic second stage, thus providing optimal conditions for nitrification.[40] The lower organic

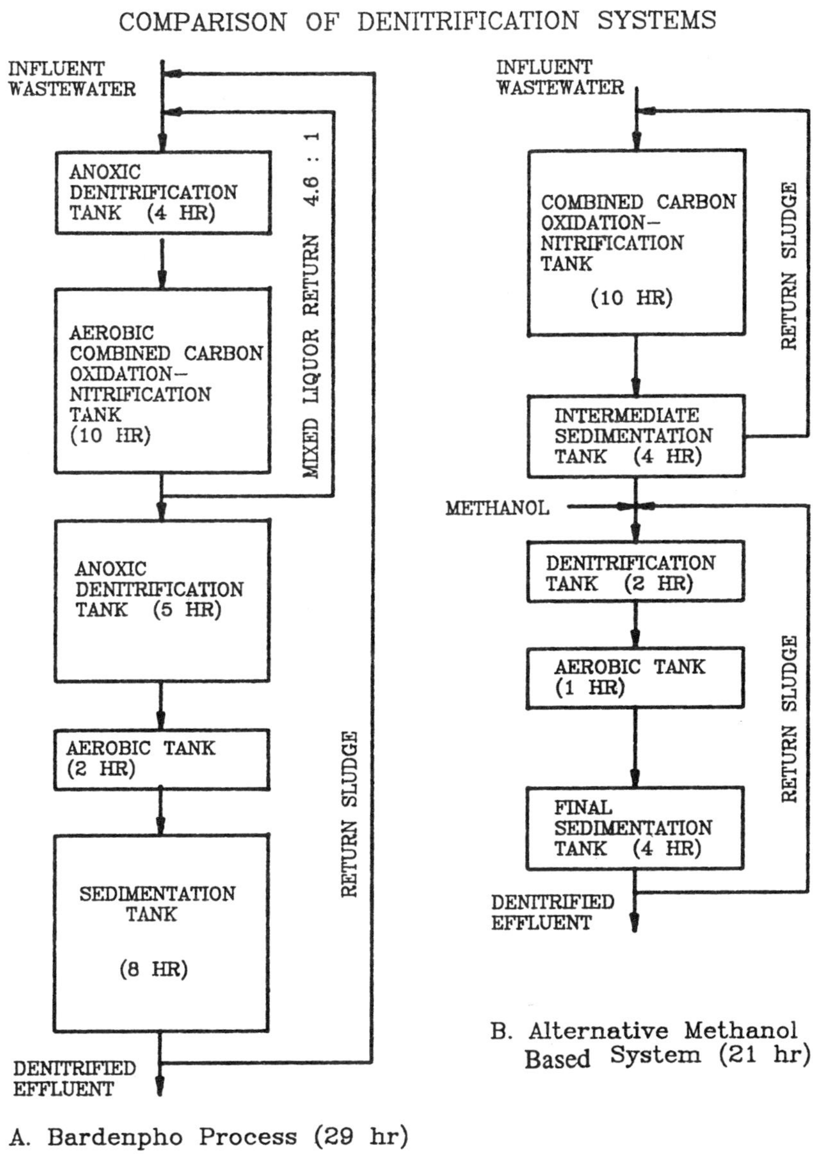

Figure 5. Comparison of denitrification systems.

loadings also result in decreased oxygen transfer requirements for the aerated basin.[39] The feasibility of one-sludge pre-denitrification systems has been demonstrated in pilot-plant[41] and full-scale studies.[42] Reduced capital costs are also realized in pre-denitrification systems since primary clarification as well as intermediate clarification may be eliminated. This savings, however, is partially offset by increased volumes required for the nitrification and denitrification basins. Another process variation utilizing the indigenous organic carbon material in wastewater for denitrification involves the utilization of alternating periods or zones of aerobic and anoxic conditions. Nitrification readily proceeds during the aerobic phase, while denitrification occurs during the anoxic phase. Oxidation of carbonaceous material will occur during both phases. According to Bishop et al.,[43] the application of such processes results in: a reduction in the volume of air required for nitrification and BOD removal; minimization or elimination of methanol for denitrification; substantial nitrogen removal achieved without special recycle of the mixed liquor; and elimination of intermediate and primary clarification. This kind of process has been successfully demonstrated in both pilot-plant[43] and full-scale[44,45] operations.

Denitrification Using Thiosulfate and Sulfide

The rising costs and increasing scarcity of methanol and similar organic compounds make it increasingly undesirable to use them as chemical additives. Thus, it is advantageous to develop denitrification processes which do not involve organic supplements.

Thiobacillus denitrificans is a species of autotrophic bacteria capable of oxidizing sulfur and sulfur compounds while reducing nitrate to free nitrogen gas. Lab-scale studies by Biscogni and Driscoll,[46] indicated that reliable autotrophic denitrification could be obtained using thiosulfate and/or sulfide as electron donors for *T. denitrificans* in a completely mixed suspended growth system. Further study in this area should be applied to a full-scale system to determine its feasibility in terms of chemical requirements and costs.

SUMMARY AND CONCLUSIONS

Significant quantities of data and a variety of process alternatives have been generated for nitrification and denitrification activated sludge systems during the past decade. The selection of an appropriate nitrification process will depend upon the following factors:

- System classification - upgrade or new.
- Permit limitations - seasonal or continuous.
- Lowest expected wastewater temperatures.
- Inclusion of phosphorous removal.
- Frequency of shock organic loads.
- Presence of inhibitors from industrial discharges.
- Land availability.
- Costs - capital, operating, and maintenance.

The selection of a denitrification process will be influenced by the kind of nitrification process chosen. If it is determined that two-stage nitrification is required then, most likely, the denitrification process will follow the nitrification stage and utilize methanol as the carbon source. Conversely, if single-stage nitrification is acceptable, the denitrification process may be incorporated into the combined carbon oxidation-nitrification reactor and utilize the organic carbon compounds present in the raw wastewater. Alternating anoxic and aerobic periods are required in this kind of combined system. A separate pre-denitrification process may also be used in conjunction with single-stage nitrification. Pre-denitrification one-sludge systems have lower oxygen, methanol, and lime requirements than for similar two-sludge post-denitrification processes and, consequently, reduced operating and capital costs. Figure 6 shows a feedforward control of methanol based on flow and nitrate nitrogen. Figure 7 shows the effect of safety factor on effluent nitrate level in suspended growth systems. Figure 8 shows the effect of diurnal variation in load on effluent nitrate level in complete mix suspended growth systems. Figure 9 shows the endless channel system for nitrogen removal. Figures 10 and 11 illustrate various nitrification systems.

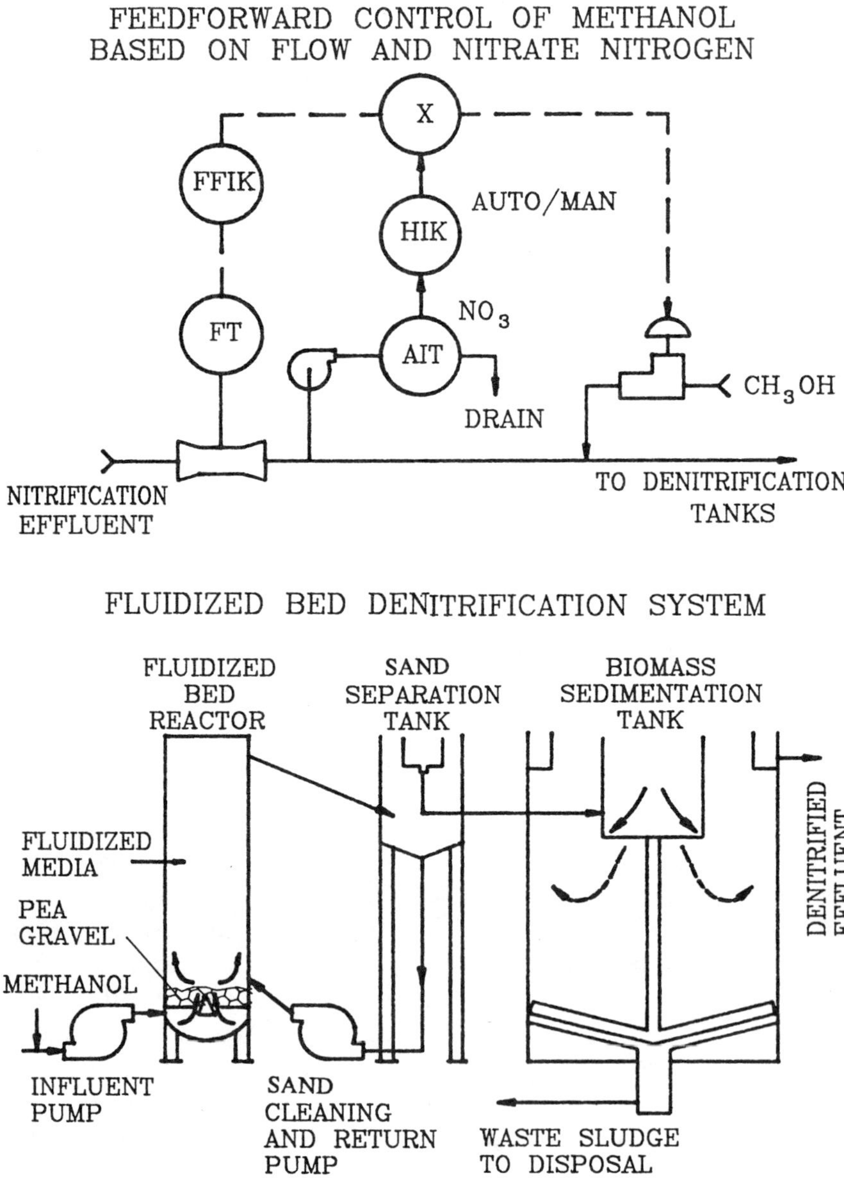

Figure 6. Feedforward control of methanol based on flow and nitrate nitrogen.

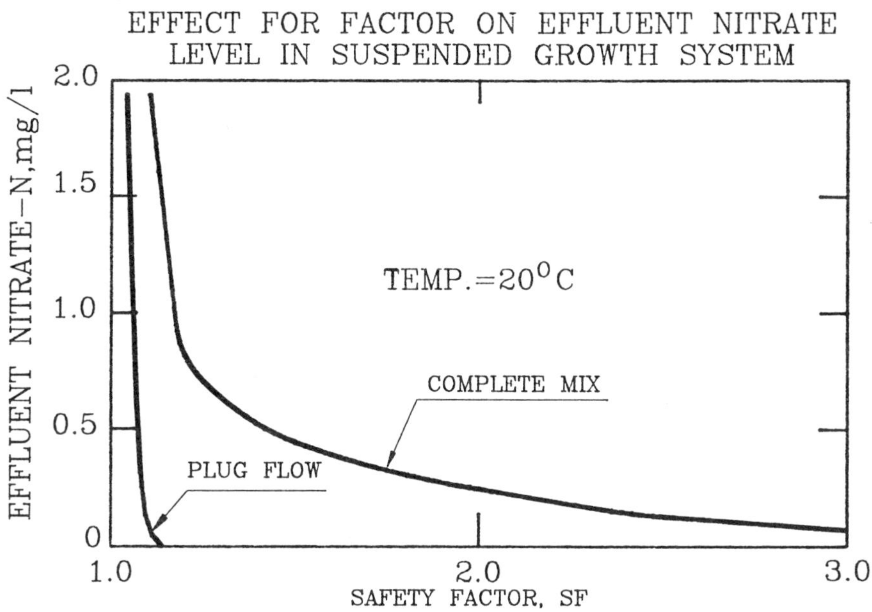

$$D_o = (\bar{D}_o - \bar{D}_l)\,\frac{\bar{Q}}{Q}\,\frac{\hat{\mu}_D}{\bar{\mu}_D}\,\frac{D_l}{K_D + D_l} + D_l$$

where:

D_o = influent nitrate -N level at any time, mg/l,

$\bar{D}_o$ = mass average influent nitrate -N level over 24 hours, mg/l,

D_l = mass average effluent nitrate -N level at any time, mg/l,

$\bar{D}_l$ = mass average effluent nitrate -N level over 24 hours, mg/l,

Q = influent flow rate at any time,

$\bar{Q}$ = average daily influent flow rate,

$\bar{\mu}_D$ = design denitrifier growth rate, day^{-1}, and

$\hat{\mu}_D$ = maximum denitrifier growth rate, day^{-1}.

Figure 7. Effect of safety factor on effluent nitrate level in suspended growth system.

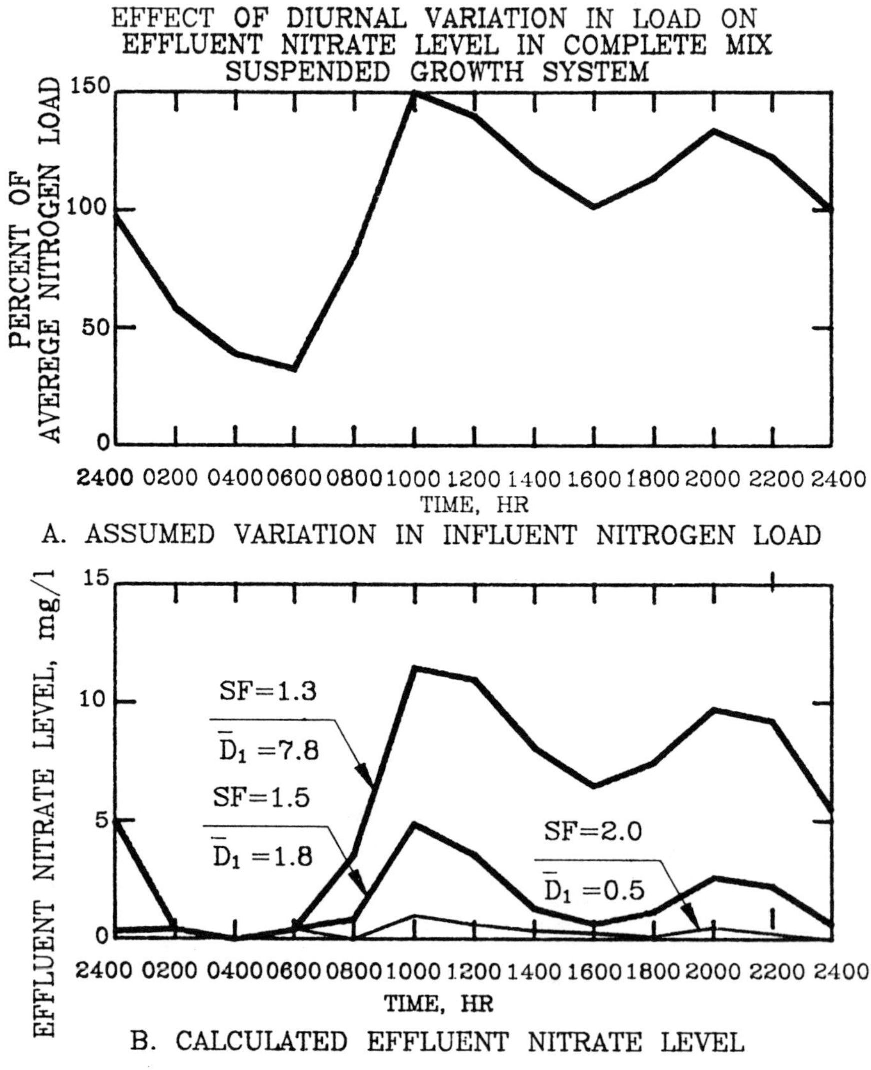

Figure 8. Effect of diurnal variation in load on effluent nitrate level in complete mix suspended growth system.

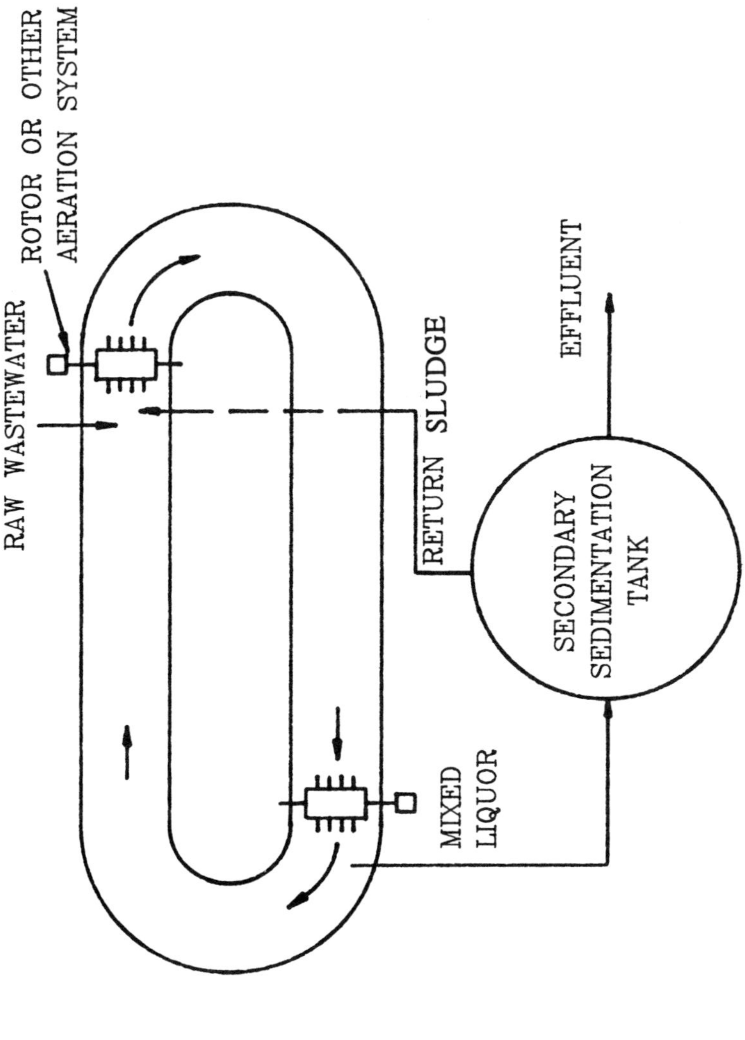

Figure 9. Pasveer ditch or endless channel system for nitrogen removal.

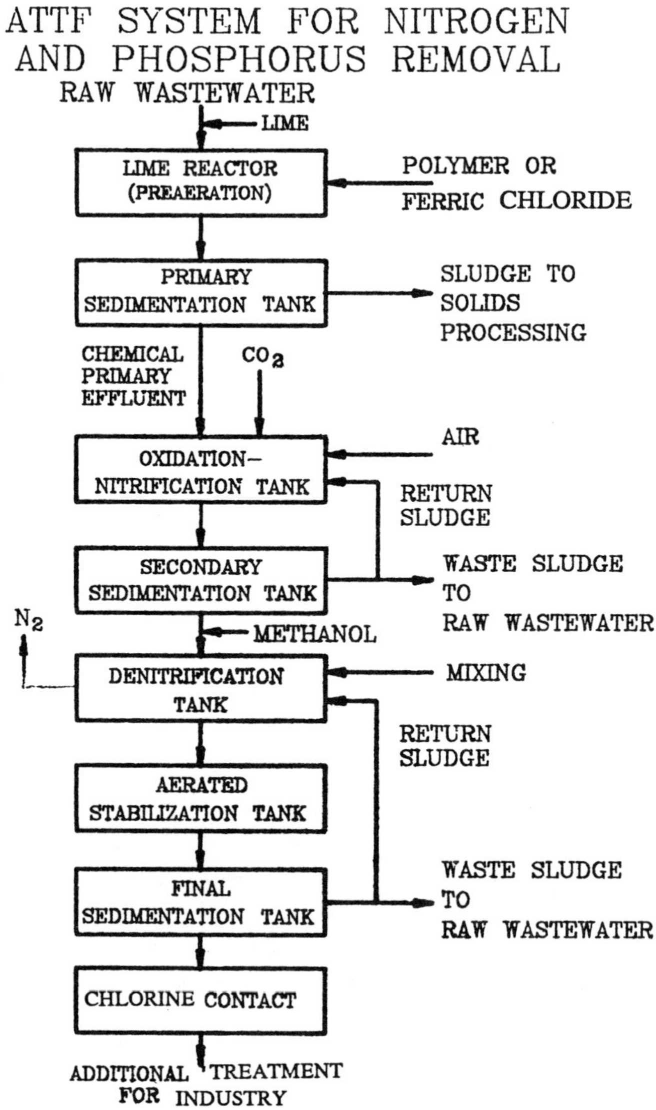

Figure 10. ATTF system for nitrogen and phosphorus removal.

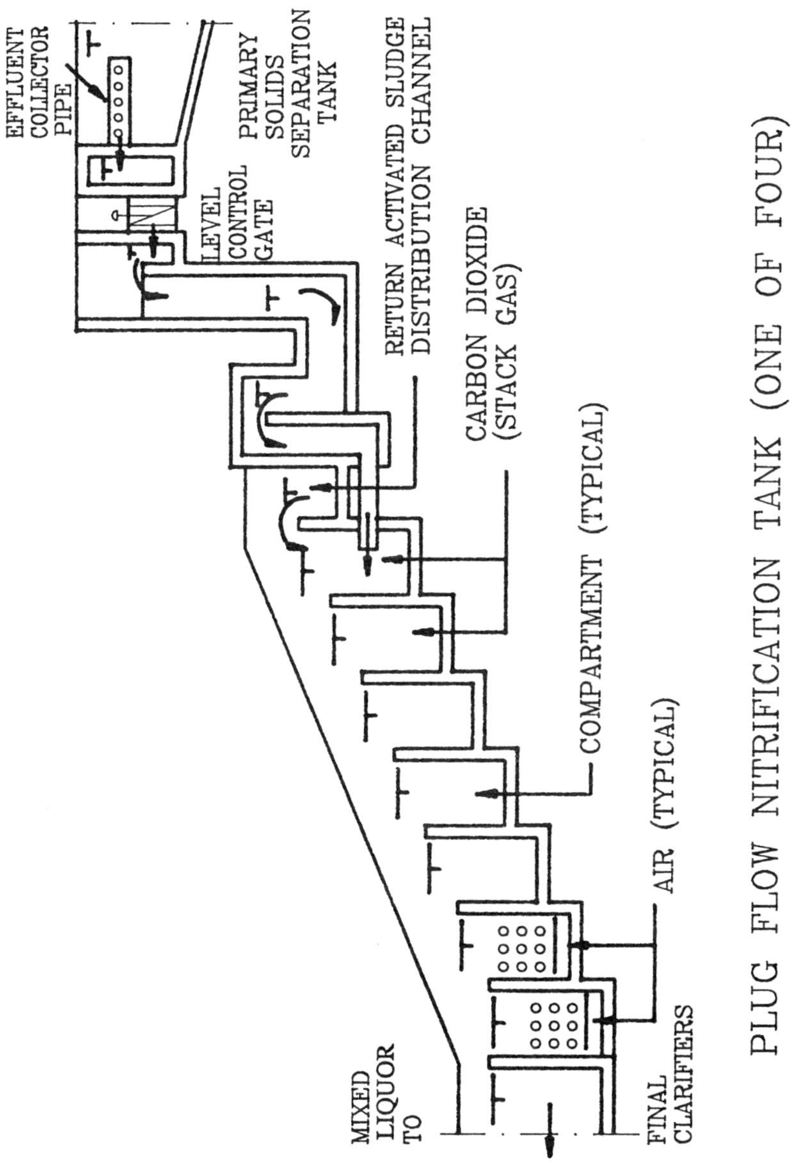

Figure 11. Section through nitrification tanks at the LMWQCC, Canberra, Australia.

REFERENCES

1. Wild, H.E., et al. "Factors Affecting Nitrification Kinetics," *Jour. Water Poll. Control Fed.*, 1971, (43), 1845.

2. Courchaine, R. J. "Significance of Nitrification in Stream Analysis - Effects on the Oxygen Balance," *Jour. Water Poll. Control Fed.*, 1968, (40), 835.

3. Sawyer, C.N., and P.L. McCarty. *Chemistry for Environmental Engineering*. New York: McGraw-Hill Inc., 1978.

4. Metcalf and Eddy, Inc. *Wastewater Engineering: Treatment, Disposal, Reuse*. New York: McGraw-Hill Inc., 1979.

5. U.S. EPA, "Process Design Manual for Nitrogen Control," Office of Technology Transfer, Wash. D.C., 1975.

6. Wang, L.K., et al. "Chemistry of Nitrification-Denitrification Process," *Jour. of Env. Sci.*, 1978, (21) #6, 23.

7. Scearce, S.N., et al. "Prediction of Alkalinity Changes in the Activated Sludge Process," *Jour. Water Poll. Control Fed.*, 1980, (52), 399.

8. Wong-Chong, G.M., and R.C. Loehr. "The Kinetics of Microbial Nitrification," *Water Research (G.B.)*, 1975, (9), 1099.

9. Poduska, R.A., and J.F. Andrews. "Dynamics of Nitrification in the Activated Sludge Process," *Jour. Water Poll. Control Fed.*, 1975, (47), 2599.

10. Sawyer, C.N., et al. "Nitrification and Denitrification Facilities," *EPA Technology Transfer Seminar Publication*, 1973, EPA-625/4-73-004a.

11. Sutton, P.M., and D.S. Parker. "Nitrification Design Approach for High Strength Ammonia Wastewaters," *Jour. Water Poll. Control Fed.*, 1978, (50), 2050.

12. Charley, R.C., et al. "Nitrification Kinetics in Activated Sludge at Various Temperatures and Dissolved Oxygen Concentrations," *Water Research (G.B.)*, 1980, (14), 1387.

13. Jenkins, D., and W.E. Garrison. "Control of Activated Sludge by Mean Cell Residence Time," *Jour. Water Poll. Control Fed.*, 1968, (40), 1905.

14. Munch, W.L. "Control Ammonia Nitrogen with Single-Stage Nitrification," *Water and Wastes Eng.*, 1978, (15) #8, 16.

15. Lawrence, A.W., and C.G. Brown. "Design and Control of Nitrifying Activated Sludge Systems," *Jour. Water Poll. Control Fed.*, 1976, (48), 645.

16. Stover, E.L., et al. "One-Versus Two-Stage Nitrification in the Activated Sludge Process," *Jour. Water Poll. Control Fed.*, 1976, (48), 645.

17. Young, J.C., et al. "Control Strategy for Biological Nitrification Systems," *Jour. Water Poll. Control Fed.*, 1979, (51), 1824.

18. Mulbarger, M.C. "Three Sludge System for Nitrogen and Phosphorous Removal," *National Technical Information Service*, PB-213 778, 1972.

19. Beckman, W.J., et al. "Combined Carbon Oxidation-Nitrification," *Jour. Water Poll. Control Fed.*, 1972, (44), 1917.

20. "Wastewater Treatment Plant Design," *Manual of Practice No. 8*, Water Pollution Control Federation, Wash. D.C., 1977.

21. Benninger, R.W., and J.H. Sherrard. "Nitrification and Alkalinity Relationships in Activated Sludge," *Jour. Water Poll. Control Fed.*, 1978, (50), 2132.

22. Stenstrom, M.K. "The Effect of Dissolved Oxygen Concentration on Nitrification," *Water Research (G.B.)*, 1980, (14), 643.

23. Hockenbury, M.R., and C. P. Grady. "Inhibition of Nitrification-Effects of Selected Organic Compounds," *Jour. Water Poll. Control Fed.*, 1977, (49), 768.

24. Joel, A.R., and C.P. Grady. "Inhibition of Nitrification-Effects of Aniline after Biodegradtion," *Jour. Water Poll. Control Fed.*, 1977, (49), 768.

25. Hockenbury, M.R., et al. "Factors Affecting Nitrification," *Jour. Env. Eng. Div.*, ASCE, 1977, (103) #EE1, 9.

26. Reinhart, D.R. "Nitrification Treatability Study for Carrollton, Ga.," *Jour. Water Poll.Control Fed.*, 1979, (51), 1032.

27. Stover, E.L., et al. "Inhibiting Nitrification in Wastewater Treatment Plants," *Water and Sew. Works*, 1976, (133) #8, 56.

28. Anthonisen, A.C., et al. "Inhibition of Nitrification by Ammonia and Nitrous Acid," *Jour. Water Poll. Control Fed.*, 1976, (48), 835.

29. Neufeld, R.D., et al. "Phenol and Free Ammonia Inhibition to *Nitrosomonas* Activity," *Water Research (G.B.)*, 1980, (14), 1695.

30. Mitchell, R. *Introduction to Environmental Microbiology*, New Jersey: Prentice-Hall, Inc., 1974.

31. Irvine, R.L., et al. "Stoichiometry and Kinetics of Biological Waste Treatment," *Jour. Water Poll. Control Fed.*, 1980, (52), 1987.

32. Mulbarger, M.C. "Nitrification and Denitrification in Activated Sludge System," *Jour. Water Poll. Control Fed.*, 1971, (43), 2059.

33. Barth, E.F., et al. "Chemical-Biological Control of Nitrogen and Phosphrous in Wastewater Effluent," *Jour. Water Poll. Control Fed.*, 1968, (40), 2040.

34. Beccari, M., et al. "A Critical Analysis of Nitrification Alternatives," *Water Research (G.B.)*, 1979, (13), 185.

35. Clarkson, R.A., et al. "Single-Sludge Pure-Oxygen Nitrification and Phosphorous Removal," *Jour. Water Poll. Control Fed.*, 1980, (52), 770.

36. Sutton, P.M., et al. "Efficacy of Biological Nitrification," *Jour. Water Poll. Control Fed.*, 1975, (47), 2655.

37. Sherrard, J.H., et al. "Utilizing Stoichiometry for Denitrification," *Water and Sew. Works*, 1978, (125) #6, 104.

38. Argaman, Y., and E. Miller. "Modeling Recycled Systems for Biological Nitrification and Denitrification," *Jour. Water Poll. Control Fed.*, 1979, (51), 749.

39. Wilson, R.W., et al. "Design and Cost Comparison of Biological Nitrogen Removal Processes," *Jour. Water Poll. Control Fed.*, 1981, (53), 1294.

40. Narkis, N., et al. "Denitrification at Various Carbon to Nitrogen Ratios," *Water Research (G.B.)*, 1979, (13), 93.

41. Sutton, P.M., et al. "Nutrient Removal in Suspended Growth Systems Without Chemical Addition," *Jour. Water Poll. Control Fed.*, 1980, (52), 98.

42. Sutton, P.M., et al. "Nitrification and Denitrification of an Industrial Wastewater," *Jour. Water Poll. Control Fed.*, 1981, (53), 176.

43. Bishop, D.F., et al. "Single-Stage Nitrification-Denitrification," *Jour. Water Poll. Control Fed.*, 1976, (48), 52.

44. Matsche', N.F., and G. Spatzierer. "Austrian Plant Knocks Out Nitrogen," *Water and Wastes-Eng.*, 1975, (12) #1, 19.

45. Schwinn, D.E., and D.F. Storrier. "One Step for Nitrogen Removal--A Giant Step for Optimum Performance," *Water and Wastes Eng.*, 1978, (15) #12, 33.

46. Biscogni, J.J., and C.T. Driscoll. "Denitrification Using Thiosulfate and Sulfide," *Jour. Env. Eng. Div.*, ASCE, 1977, (103) #EE4, 593.

5 IN-SITU BIORECLAMATION OF CONTAMINATED GROUNDWATER

INTRODUCTION

In-situ bioreclamation of contaminated groundwater is a promising new technique for enhancing the clean-up rate of aquifers contaminated with organic pollutants, such as chlorinated solvents, gasoline constituents, and pesticides. In-situ bioreclamation involves injecting the materials necessary to significantly increase the microbiological activity in the subsurface. The injected material is a component that limits the growth of the desired microorganisms and is usually an electron acceptor (e.g., oxygen or nitrate), a carbon source, or a macro-nutrient (e.g., nitrogen or phosphorus). Injecting the proper amount of the limiting material creates a region of increased microbiological activity, called the Biologically Active Zone (BAZ).

Creation of a BAZ offers major advantages for aquifer clean-up because the removal agents, the bacteria, are in close proximity to all the contaminants, including those in the water, those sorbed to aquifer materials, and those in a nonaqueous liquid phase. Thus, the relatively slow mechanism of flushing by water flow is replaced by a degradation reaction very near the source of contaminants.

A study by Rittmann et al.[1] centered around an increased ability to understand and quantitatively describe the key phenomena affecting the formation of and reactions within a BAZ. The specific objectives of the study were to:

1. Develop a laboratory-scale, porous-medium column that can be used to create and study BAZs under well-defined conditions.

2. Evaluate the formation of one or more BAZs within the laboratory-scale column when the electron acceptor, nitrate, is injected into the flow path.

3. Using the laboratory-scale columns, evaluate the fate of commonly found halogenated solvents as they passed through the BAZs.

4. Develop and test an efficient computer model for the formation of BAZs and the utilization of substrates by the BAZs.

5. Apply the model to describe and interpret the formation of the BAZ and the fates of the various substrates in the column experiments.

6. Employ the model to evaluate strategies for in-situ bioreclamation in the field.

The laboratory-scale, porous-medium columns were constructed of 2.5-cm inside diameter by 22.5-cm long glass tubes and were filled with 3-mm glass beads. Ports were placed every 2.5 cm to allow for sampling and/or injection. Special injection assemblies were designed to allow for uniform planar injection of substrate into the flow path. The systems gave an excellent approximation of one-dimensional flow.

The organic source was sodium acetate, which was available as a C-radio-labelled tracer.[14] It was fed continuously to the inlet end of the column from an elevated reservoir. In most experiments, the injected material was the electron acceptor, NO_3^-. One or two BAZs were established at and downstream of the injection ports. In order to ensure that NO_3^- was limiting, no other electron acceptors were added to the feed medium, and extreme measures were taken to preclude O_2 entry in the reservoir, feed lines, and columns.

Well-defined BAZs developed from the injection ports and up to 7.5 cm downstream of the injection ports. Photography of the intact columns and of the beads in the columns demonstrated that the bacterial growth was present as biofilms on the glass beads. Photography and measurements of biofilm mass on the beads confirmed that the amount of accumulated biofilm was greatest right after the injection port, and it gradually declined downstream.

Acetate (expressed as soluble organic carbon, SOC) and NO_3^- declined across the BAZs according to the expected stoichiometry, 0.67 mg NO_3^- -N/mg SOC. For the column with two BAZs, removal of SOC was partial in the BAZ after the first injection port, because NO_3^- was depleted; however, SOC removal was essentially complete in the second BAZ, as sufficient NO_3^- was supplied in the second injection. These results demonstrated that stoichiometric addition of an electron acceptor could be used to remove an electron-donor substrate to the degree desired.

Formation of N_2 gas bubbles occurred as a result of the denitrification of NO_3^-. These bubbles tended to accumulate in the BAZs and caused some short-circuiting, which led to a deterioration of SOC removal. Removal of the bubbles restored the SOC removal and demonstrated the possible deleterious effects of gas evolution within a confined aquifer.

Eight trace-concentration halogenated solvents were applied to the feed of the column having one BAZ. Two dichlorobenzenes were added together as a mixture, and six one- or two-carbon halogenated aliphatics were added as another mixture. Of the halogenated aliphatics, carbon tetrachloride was removed the most completely by the denitrifying BAZ. Tetrachloroethane, bromoform, ethylene dibromide, and trichloroethene were removed to lower degrees. Trichloroethane was slightly removed. 1,2 and 1,3-dichlorobenzene also were 20-30% removed during passage through the BAZ. Significant increases in the fractional removal were effected as the liquid flow velocity was decreased, which increased the contact time in the BAZ. These results are especially significant for two reasons. First, they show that common groundwater contaminants were degradable in the BAZs induced by NO_3^- injection. Most interesting are the removals of the dichlorobenzenes and trichloroethene, which were thought previously to be refractory under denitrifying conditions. Second, the results show that the removals of each compound depended upon the degradation kinetics of the particular compound and the contact time in the BAZ.

Modeling of the formation of a BAZ was based on application of biofilm kinetics to solute transport in porous media. A steady-state-biofilm model was incorporated into a one-dimensional, steady-state, solute-transport equation. The equation was transformed from the differential form to one using discrete finite differences and solved numerically directly for the steady-state profiles of substrate concentration and biofilm accumulation. Major modeling advancements were the ability to have lateral injection sources at any point along the column and the use of quasilinearization to give a highly efficient and direct solution for the steady state. The quasilinearization technique, which involves substituting a first-order Taylor series approximation for the highly nonlinear reaction term, made the convergence to steady state approximately ten times faster than by conventional methods. Even greater improvements are expected for more complicated geometries.

The modeling also was advanced by explicit coupling of the steady-state-biofilm model solution, which solves for the concentration profile of the limiting substrate and the amount of biofilm, to models for a non-limiting substrate and for secondary substrates. An example of a non-limiting substrate is NO_3^- when SOC is limiting; the flux of NO_3^- into the biofilm was set equal to the flux of SOC multiplied by a stoichiometric coefficient. Although the flux of the non-limiting substrate was determined by the flux of limiting substrate, it had its own rates of advection, dispersion, and injection. A secondary substrate is, in this context, a trace-level contaminant that is removed in the BAZ, even though its utilization provides negligible or no benefit to the microorganisms. The flux of secondary substrate was determined by its own kinetic characteristics and by the amount of biofilm accumulated through utilization of the SOC.

The steady-state, solute-transport model for the limiting substrate and the coupled transport model for the non-limiting substrate were used to evaluate the experiments on the formation of BAZs. Kinetic parameters for the utilization of the SOC were determined independently; thus, model results were true predictions. SOC, NO_3^-, and biofilm profiles matched the experimental results very well for columns with one and two BAZs. Model predictions and experimental results agreed quantitatively that removals of SOC and NO_3^- and accumulation of biofilm were greatest in the first 2.5 cm beyond the injection port. Removal rates and biofilm accumulation declined gradually in the next 5.0 cm, and substrate concentrations attained a steady plateau value thereafter. Predictions and experimental results also concurred that injection of more of the limiting material (NO_3^- in this case) allowed formation of a second BAZ and renewed removal of SOC. The model predictions correctly described all trends, and absolute deviations between predicted and experimental results were small in all cases.

The coupled transport model for secondary substrates also was used successfully for describing the removal of the halogenated aliphatic solvents. Since the kinetic parameters for each secondary substrate could not be determined independently, one set of results from the column experiments was used to obtain a best-fit set of kinetic parameters. These parameters were then used to predict the removal across the BAZ for experiments with different liquid flow velocities. Model and experimental results agreed well on the effect of liquid flow velocity. When the flow velocity was decreased, the contact time for the secondary substrates in the BAZ was increased proportionally. This increase in contact time allowed greater removal. For example, experimental and modeling results agreed that the removal of carbon tetrachloride through a BAZ should increase from 18% to 55% to 92% as the post-injection detention time increased from 50 minutes to 125 minutes to 500 minutes, respectively.

The steady-state models were applied to investigate possible strategies to be used in field bioreclamations. The use of multiple injection wells was studied for its ability to decrease aquifer clogging potential by spreading out the distance over which limiting substrate is added. Modeling results verified that the strategy of multiple injections could reduce high densities of biofilm accumulation near the injection well. Also investigated was the strategy of adding a supplemental carbon source to extend the length of a BAZ. The modeling illustrated that such an extension of the BAZ could be accomplished and could result in longer contact times of a secondary substrate with the BAZ, thereby increasing the removal of the secondary substrate.

TREATING CONTAMINATED GROUNDWATER

Contamination of groundwater by organic materials--such as chlorinated solvents, petroleum products, and landfill leachates--is widely recognized as one of the most critical environmental problems of recent times. Currently, clean-up efforts usually involve extraction of the contaminated water, followed by physical, chemical, or biological treatment. Because the organic contaminants can adsorb onto aquifer solids or can be trapped in regions of relatively low permeability, the volume of water required to be extracted is many times larger than the pore volume that is contaminated; thus conventional clean-up is very expensive and time-consuming.

In-situ biological degradation is being proposed as a promising alternative for aquifer restoration. In-situ projects typically involve a set of extraction and injection wells, which establishes a defined flow field and permits inputs of seed microorganisms, electron acceptor, carbon source, or other nutrients at one or more points along the flow path. Being a very new technology, in-situ bioreclamation designs have been based on only a few simple microbiological experiments aimed at testing biodegradation potential and nutrient requirements. Incorporation of realistic biodegradation kinetics and groundwater hydraulics has not been accomplished.

An initial requirement of any in-situ decontamination technology is that the flow field be defined. Otherwise, contaminants can escape treatment by migrating out of the treatment site or by remaining in isolated portions of the aquifer. For in-situ bioreclamation, however, more than a defined flow field is required: the water in that flow field must pass through a biologically active zone before it is extracted or leaves the treatment site. The biologically active zone in an aquifer is made up almost completely by microorganisms attached as biofilms to the large amount of surface area presented by the aquifer solids. Even in uncontaminated aquifers, bacteria are found attached to aquifer solids; however, their densities are very low (10^6/gram of soil), and their metabolic capabilities are largely undefined. Successful in-situ bioreclamation requires that the attached biomass be increased greatly from that normally found on aquifer solids. In some cases, different types of microorganisms, having capabilities not found in the natural community, should be added as seed. In almost every situation, however, success requires that the microorganisms grow to attached densities a hundred or more times that naturally present.

Cell growth and accumulation in an aquifer depend on the availability of an electron donor, an electron acceptor, and several other nutrients, such as nitrogen, phosphorous, and sulfur. Usually, one factor is rate limiting which controls how much cell mass can be accumulated. The growth-limiting factor can be called the limiting substrate. Enhanced in-situ bioreclamation usually involves adding the limiting substrate in such a manner that the growth limitation is eliminated and significant quantities of biomass are generated in the aquifer.

What the limiting substrate must be varies with the contaminating situation. For instance, a leak or spill that creates high organic-contaminant concentrations probably is limited by the electron acceptor or a nutrient. On the other hand, low-level contamination of a drinking water supply by a distant source creates a situation in which an organic electron donor is needed to allow significant growth.

The objective of enhanced in-situ bioreclamation is to establish a biologically active zone by supplying the limiting substrate in such a manner that no contaminant escapes biodegradation. However, biodegradable material added via an injection well to enhance in-situ biodegradation often is consumed very rapidly near the injection well, creating two significant problems: (1) biological growth is limited to only a region very near the injection well, and (2) well clogging can occur. The first problem is quite serious, since localization of biological activity prevents adequate contaminant/microorganism contact throughout most of the aquifer. The second problem also is serious because clogging retards the input of the limiting substrate and may force the groundwater flow to go around the biologically active zone.

The problem of localized biological activity can be solved, at least in principle, by providing multiple injection wells perpendicular to and/or along the flow path. Figure 1(a) depicts the case where multiple injection wells are placed laterally along the flow path to create a saw-tooth pattern of nutrient concentration, which allows the biological activity to extend the necessary distance to assure adequate contaminant removal. Figure 1(a) also demonstrates the concept of *coupled* hydraulic/biological reclamation, since a network of injection/extraction wells is utilized to hydraulically isolate the contaminant plume from the natural groundwater flow regime. To date, all reported cases of in-situ bioreclamation have included hydraulic control measures; in fact, several projects did not utilize multiple injection wells along the flow path and, thus, biological activity was most likely concentrated in the vicinity of the hydraulic-control injection wells.

The need for multiple injection wells along the flow path is most acute for two commonly encountered situations. The first occurs when the limiting substrate is oxygen, a common electron acceptor. Because of the low solubility of dissolved oxygen (about 9 mg/l when exposed to the atmosphere) and its reactivity with reduced materials, supplying dissolved oxygen from one injection point cannot maintain sufficient dissolved oxygen throughout the flow path when the amount of organic material to be degraded is more than a few mg/l. Since degradation of certain common classes of compounds, especially including benzene derivatives, appears to occur best (and likely exclusively) when oxygen is available, the application of oxygen is likely to be a major vehicle for enhanced in-situ bioreclamation. Other oxygen sources are ozone and hydrogen peroxide; although application of these materials overcomes some of the solubility problems of dissolved oxygen, they are reactive with reduced materials and are toxicants to microorganisms. Thus, they cannot be applied in unlimited amounts.

The second common occasion when multiple injections are needed along the flow path occurs when an electron donor, usually an organic compound, must be applied to allow increased growth of microorganisms that bring about contaminant removal through secondary utilization or co-metabolism. Because the electron donor can be utilized quickly near the injection well, small input concentrations do not penetrate far into the aquifer, but large concentrations cause well clogging through biomass plugging or gas binding (in methanogenic or denitrifying cases). Thus, the electron donor input must be spread out along the flow path to give a sufficient amount of microorganisms without plugging the aquifer.

Figure 1(b) shows the case in which multiple injection wells are placed perpendicular to the groundwater flow path to create a biologically active zone through which all of a contaminant plume must pass. This bioreclamation scheme is probably less expensive than that shown in Figure 1(a) since hydraulic control measures are not utilized. Creating a biologically active zone perpendicular to the natural groundwater flow path is a novel concept in the field of aquifer restoration.

APPLICATION OF MODELING

SOC and NO_3^- Profiles

The one-dimensional model developed by Rittmann et al.[1] was evaluated for its ability to describe the laboratory results on SOC and NO_3^- removal through the BAZs. The assumptions used for applying the one-dimensional solute transport equation to the laboratory column are that wall effects were negligible and the surface area due to the sides of the

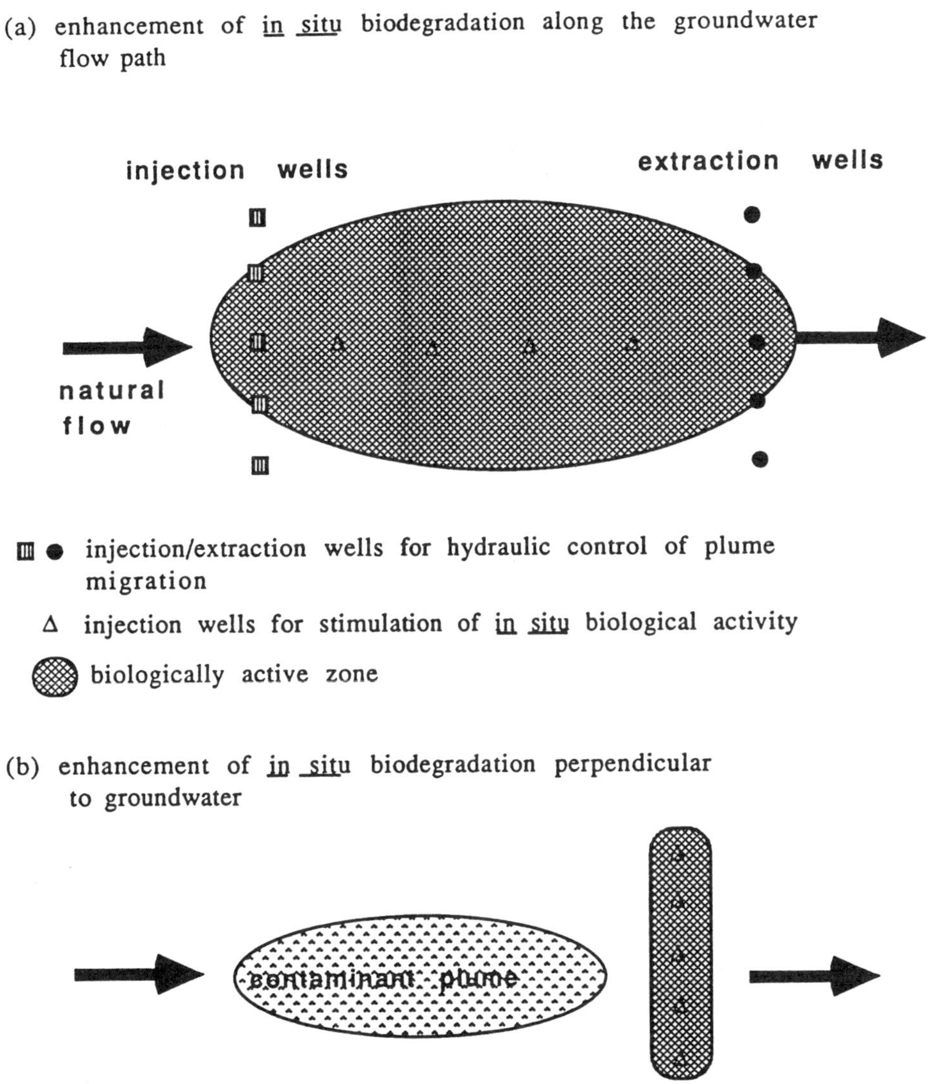

Figure 1. Strategies for in-situ bioreclamation of contaminated groundwater.

column (less than one percent of the area of the glass beads) need not be included.

The kinetic parameters (k, K_s, Y, b, and X_f) were determined independently. The kinetic and reactor parameters used to model the laboratory results are presented in Table 1. The value of S_{min} determined from the kinetic parameters is slightly less than the plateau concentration of SOC measured in the laboratory. This can be explained by the formation of soluble microbial products (SMP) (Namkung and Rittmann, 1986) which contained C^{14}. Thus, SOC measurements toward the downstream end of the column contained residual substrate and some SMP, while the model predictions are only for residual substrate.

One-BAZ Column

The modeling procedure was a two-step process. First the SOC profile was solved by assuming SOC was the rate-limiting substrate. This yielded steady-state profiles of SOC concentration and J_{SOC}, the flux of SOC into the biofilm. Then, the NO_3^- profile was obtained by solving

TABLE 1

PARAMETERS USED IN SOLUTE-TRANSPORT MODELING OF ONE-BAZ COLUMN

Parameter	Units	Value Used in Model	Value with Alternative Units
Acetate	as SOC	-	
S^o	mgSOC/l	6.5	
L	cm	0.02195	
S_{min}	mgSOC/l	0.0131	
k	mgSOC/mgcell-day	2.22	4.17 mg SOC/mgcellC-d
K_s	mgSOC/l	0.218	
X_f	mg cells/cm^3	15.	8.0 mg cell C/cm^3
Y	mg cells/mg SOC	0.678	0.36 mg cell C/mgSOC
b	day^{-1}	0.07	
D_{SOC}	cm^2/day	1.07	
D_H	cm^2/day	120.58	
v	cm/day	144.	
a	cm^{-1}	20.0	
ϵ	cm^3/cm^3	0.30	
NO_3^-	as N	-	
S^o	mg NO_3^- -N/l	7.32	
D_N	cm^2/day	1.40	

the solute-transport equation for NO_3^- when the rate of NO_3^- removal was equal to the flux of SOC multiplied by a stoichiometric coefficient. The stoichiometry was found in the laboratory to be 0.67 mg NO_3^- -N/mg acetate as SOC. That is, equation (4.2) was solved for the NO_3^- concentration profile with $J_{NO3} = 0.67\ J_{SOC}$. The numerical values for flux of NO_3^- were computed from stored values of the SOC obtained with the primary substrate model and were calculated for each grid point. Since the rates of NO_3^- were determined by multiplying the flux of SOC at each grid point by 0.67 mg NO_3^-/mg SOC, the governing transport equation was linear, so that quasilinearization was not required. In other words, dJ/dS was zero for NO_3^-, because J was a predetermined constant.

The model results are compared with the experimental results in Figure 2. The model and laboratory results compare very well. Both substrates were removed rapidly in the first 5.0 cm downstream of the injection. They then approached a plateau concentration beyond about 10 cm, as the SOC primary substrate approached its S_{min}. The correspondence between model and experimental results for both substrates verifies that SOC was rate limiting and that the stoichiometry between NO_3^- and SOC removals was correct. While there is nearly perfect agreement for the electron acceptor, NO_3^-, small deviations for SOC occur at 5.0 and 7.5 cm downstream of the injection port. These deviations may be the result of short circuiting due to nitrogen gas build-up, or they may be caused by sampling error.

Two-BAZ Column

The two-BAZ column was modeled using the solute transport model and the same reactor and kinetic parameters as for the one-BAZ column. The same influent SOC concentration was used as in the one-BAZ experiment; however, the electron acceptor was injected in two locations, the second injection being ten centimeters downstream from the first. The same total amount of electron acceptor was injected into both columns, but the two-BAZ column received 25% through the first port and the remaining 75% through the second port; this corresponds to 1.92 and 5.52 mg NO_3^- -N/l respectively. This two-injection strategy caused NO_3^- to be the rate limiting substrate in the first BAZ, where it was depleted to close to its S_{min} just before the second injection. At this point, there was approximately 50% removal of acetate. After the second injection, NO_3^- was in ample supply, and SOC (acetate) became the rate-limiting substrate.

The change of rate-limiting substrate after the second injection of NO_3^- presented an interesting modeling situation. If the electron acceptor had limited the growth throughout the length of the column, the modified model with lateral injection ports could have been used directly. In the case of a change of limitation, however, two coupled solute-transport equations had to be used.

In the section of the column after the first injection, quasilinearization and finite differences were used to solve the solute transport equation for NO_3^-, the rate limiting substrate. Then, the profile for SOC was obtained from the NO_3^- fluxes and stoichiometry, as reported in the previous section. At the point of the second injection of nitrate a new solute transport equation had to be solved. For the points downstream of the second injection, this new solute transport equation was solved using SOC as the primary substrate; it was coupled to the upstream segment of the column by considering the continuity of SOC flux at the injection port. For NO_3^-, the upstream flux of NO_3^- was added to the flux through the injection port, as it represented only approximately 0.18% of the flux through the port. The NO_3^- profile after the second injection was obtained from the SOC fluxes and stoichiometry.

Table 2 shows the parameters used in the solute transport modeling. The kinetic parameters for NO_3^- were not independently measured in the laboratory and had to be estimated. The maximum specific rate of substrate utilization, k, was taken from the k of SOC, adjusted by stoichiometry. The K_s value was varied until proper fit of the laboratory data was obtained. The low value for K_s for NO_3^- is consistent for electron acceptors (Rittmann and Langeland, 1985). The kinetic parameters for SOC, the primary substrate after the second NO_3^- injection, were averaged from those measured independently at different locations in the column.

Figure 3 shows the numerical results compared to the laboratory data. The numerical results are in extremely good agreement with the laboratory data. The stoichiometric values used in the numerical work, 1.5 mg SOC/mg NO_3^- and 0.67 mg NO_3^-/mg SOC for the first and second BAZ, respectively, allowed proper representation of both substrate profiles in both BAZs. Thus, the choice of which substrate was rate-limiting seems justified.

Comparison of Figures 2 and 3 demonstrates that having two NO_3^- injections spread out the distance over which a BAZ was present. With two injections, the BAZ covered about 12.5 cm, while it covered about 7.5 cm for one injection.

TABLE 2

PARAMETERS USED IN SOLUTE-TRANSPORT MODELING OF TWO-BAZ COLUMN

Parameter	Units	Value
Nitrate	as NO_3^- -N	-
S^o	mg NO_3^-/l	1.92, 5.52
L	cm	0.02195
$S_{min\text{-}N}$	mg NO_3^-/l	0.0090
k	mg NO_3^-/mg cell-day	1.45
K_s	mg NO_3^-/l	0.146
X_f	mg cells/cm^3	15.
Y_N	mg cells/mg NO_3^-	1.02
b	day^{-1}	0.07
D_N	cm^2/day	1.07
D_{HN}	cm^2/day	120.58
v	cm/day	144.
ϵ	cm^3/cm^3	0.30
Acetate	as SOC	-
S^o	mg SOC/l	7.09
k	mgSOC/mgcell-day	2.00
K_s	mgSOC/l	0.80
X_f	mg cells/cm^3	15.
Y	mg cells/mg SOC	0.678
S_{min}	mgSOC/l	0.0497
D	cm^2/day	1.40

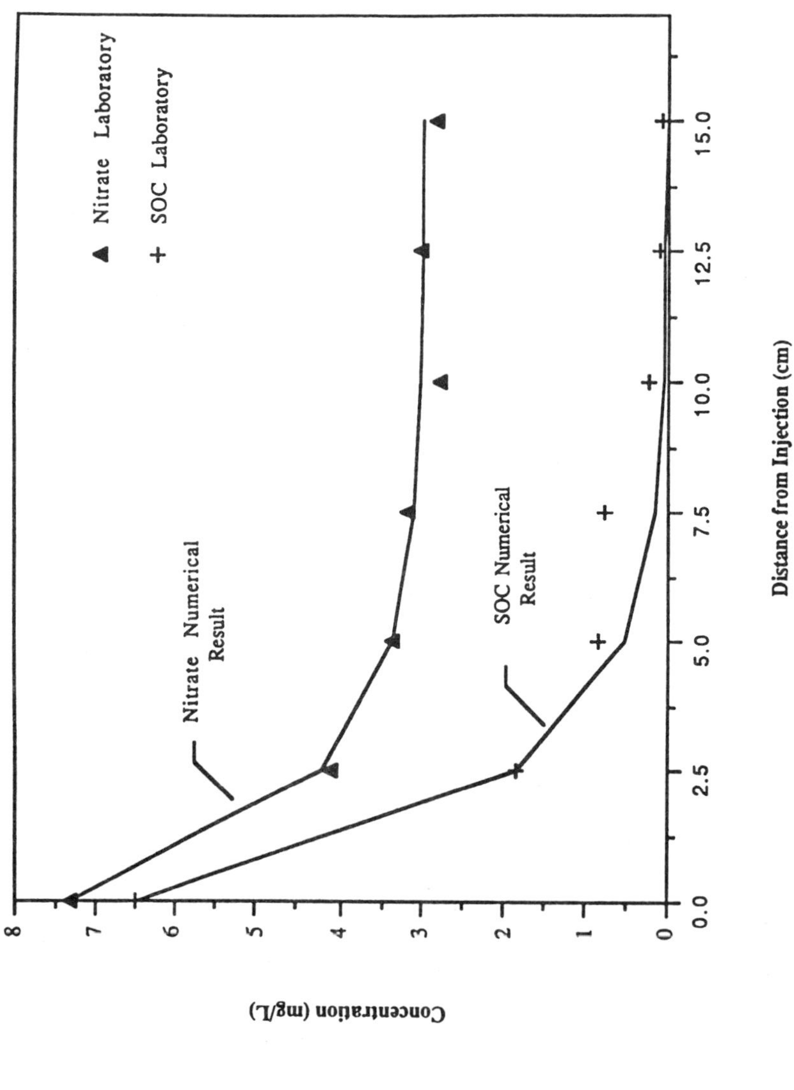

Figure 2. Comparison of laboratory and numerical results for the one-BAZ column. Zero distance indicates the injection port.

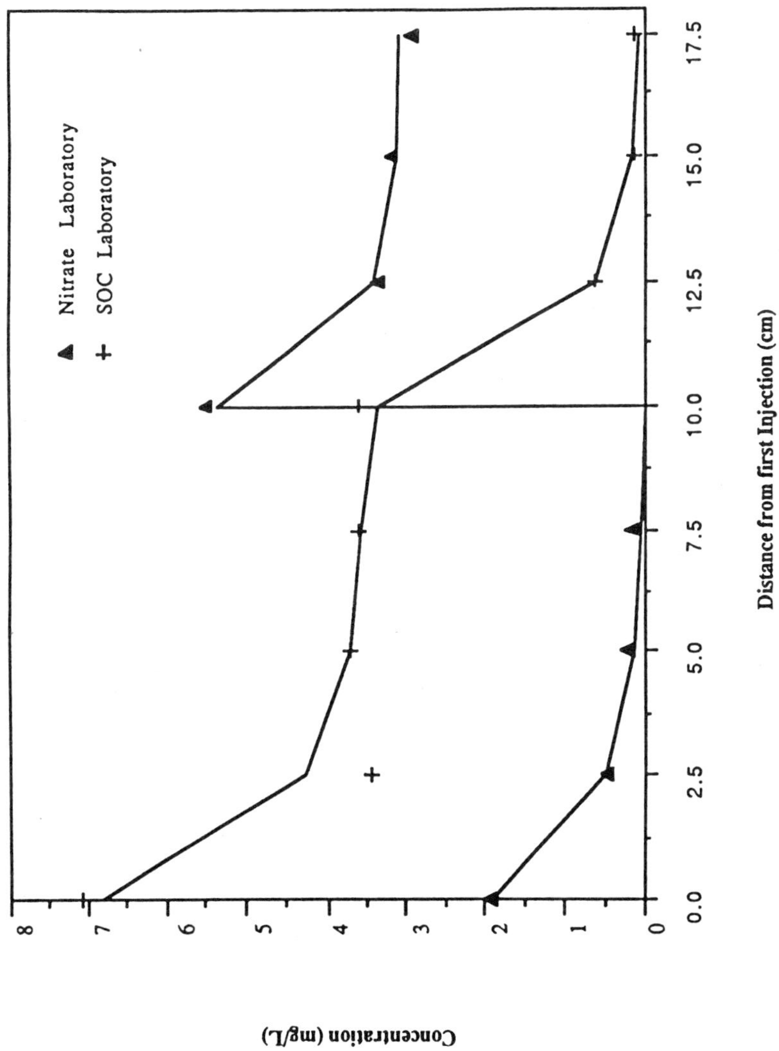

Figure 3. Comparison of laboratory and numerical results for the two-BAZ column. Nitrate injections are at 0.0 and 10.0 cm. Lines represent model prediction.

Secondary Substrate Profiles

The primary substrate profile was modeled first. From the results for the primary substrate, the steady-state biofilm thickness was calculated at each grid point from the following equation (Rittmann and McCarty, 1980a)

$$L_{fi} = \frac{J_i Y}{b X_f} \tag{1}$$

where J_i is the flux into the biofilm at grid point i. The result was a profile of biofilm thickness throughout the length of the column. The L_f values then served as key inputs to the model to estimate the flux of a particular secondary substrate into the biofilm (Rittmann and McCarty, 1981; Namkung et al., 1983).

The strategy for modeling the secondary substrates was to choose k and K_s values that provided a good fit to the experimental results for one experiment. This fitting exercise was needed, because the k and K_s values of these secondary substrates were not known independently for the denitrification system. The appropriate reactor parameters and the kinetic parameters of the primary substrate were known independently. The fitted k and K_s values were used to predict the results for different experiments with the same secondary substrates.

Several secondary substrates were fed continuously into the laboratory column. The original detention time of 50 minutes showed only slight removal of one compound, CTC. To obtain better removals, the detention time was increased at first to 125 minutes and then to 500 minutes. Most of the compounds showed significantly greater percentage removal as the detention time increased. Three categories of compounds resulted: (1) CTC was rapidly removed, (2) BF, EDB, TeCE, and TCE were removed less rapidly, and (3) TCA was not removed. To evaluate the secondary-substrate modeling, CTC, BF, EDB, TeCE, and TCE were modeled.

Carbon Tetrachloride

The first secondary substrate modeled was CTC, which entered the first BAZ at a concentration of 81.0 μg/l. The k and K_s values which gave a good fit for a detention time of 50 minutes were 0.030 μg/mg cell-day and 4.5 μg/l, respectively. Figure 4 shows the results of the numerical fitting for the 50-min. detention time. Clearly, all of the points are well represented by the numerical result. In order to demonstrate the predictive ability of the numerical model, the profiles at the two other detention times (125 and 500 min.) were calculated using the same k and K_s values. The superficial velocity and influent concentration are suitably modified for each of the other runs.

The 125-min. detention time was modeled by changing the superficial velocity and influent concentration to 0.04 cm/min. and 69 μg/l, respectively. Figure 5 shows the results of the numerical prediction compared to the laboratory data. The two curves fit well and show removal through all of the BAZ (~ 7.5 cm). The 500-min. detention time column was modeled using the laboratory obtained concentration of 53 μg/l and an adjusted superficial velocity of 0.01 cm/min. Figure 6 shows the results of the numerical prediction. The results are encouraging, because the numerical and experimental results have the same trends of rapid decrease and approach a plateau concentration. The absolute values of the plateau concentration are slightly different, but are in the same order of magnitude.

Before the secondary-substrate experiments were started, the one-BAZ column had been operated for over one year at approximately the same detention time and concentration of substrates. By the time the secondary substrates were added to the column, a steady-state biomass distribution had been attained throughout the length of the BAZ. One possibility is that the biomass distribution changed during the experiments at longer detention times and, to some degree, with different influent SOC concentrations. Reduction of the detention time reduced the substrate (SOC) input to the BAZ and should have created a situation of slower biological activity in the BAZ. The time that the new detention times were maintained before sampling of the profiles of the secondary substrates is a major factor in the assumption made about the biomass distribution for the secondary-substrate measurements. Because of the long time used to establish the steady-state biofilm and the relatively short times the reactors were run at the new detention times (6 weeks at 125 min. and 5 weeks at 500 min.), it was assumed that the biomass distribution remained the same as that calculated for the 50 minute detention time. Nevertheless, it is possible that the BAZ lost active biomass during the detention-time experiments. The data shown in Figures 4-6 suggest that the loss of activity for CTC removal was negligible.

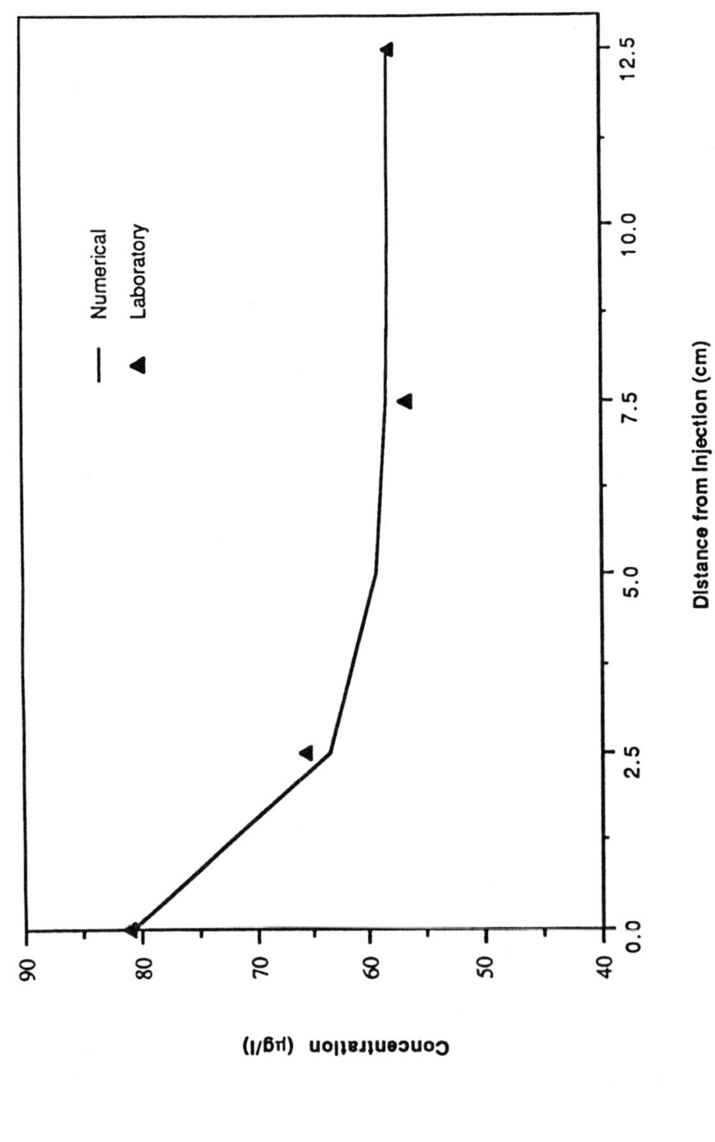

Figure 4. Numerical curve fit to the CTC profile at a detention time of 50.0 min. The k and K_s are 0.030 µg/mg cell-day and 4.5 µg/l, respectively.

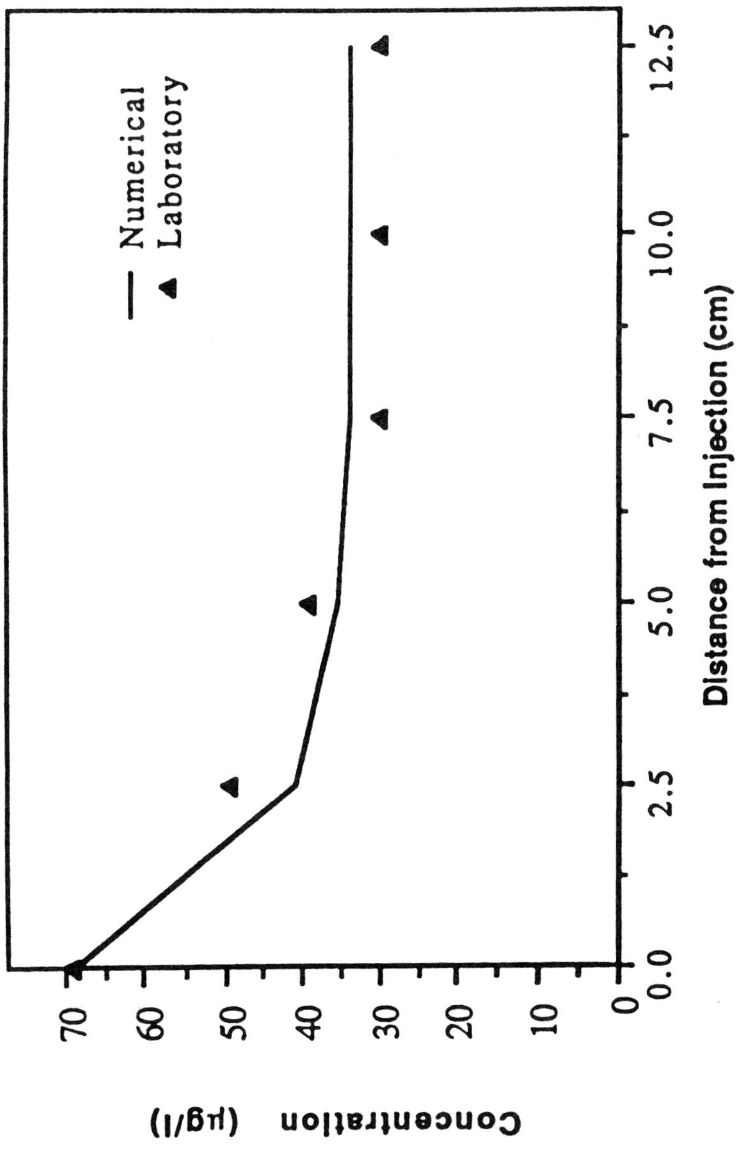

Figure 5. Prediction of the CTC profile at a detention time of 125 min. and with $k = 0.030$ µg/mg cell-day and $K_s = 4.5$ µg/l.

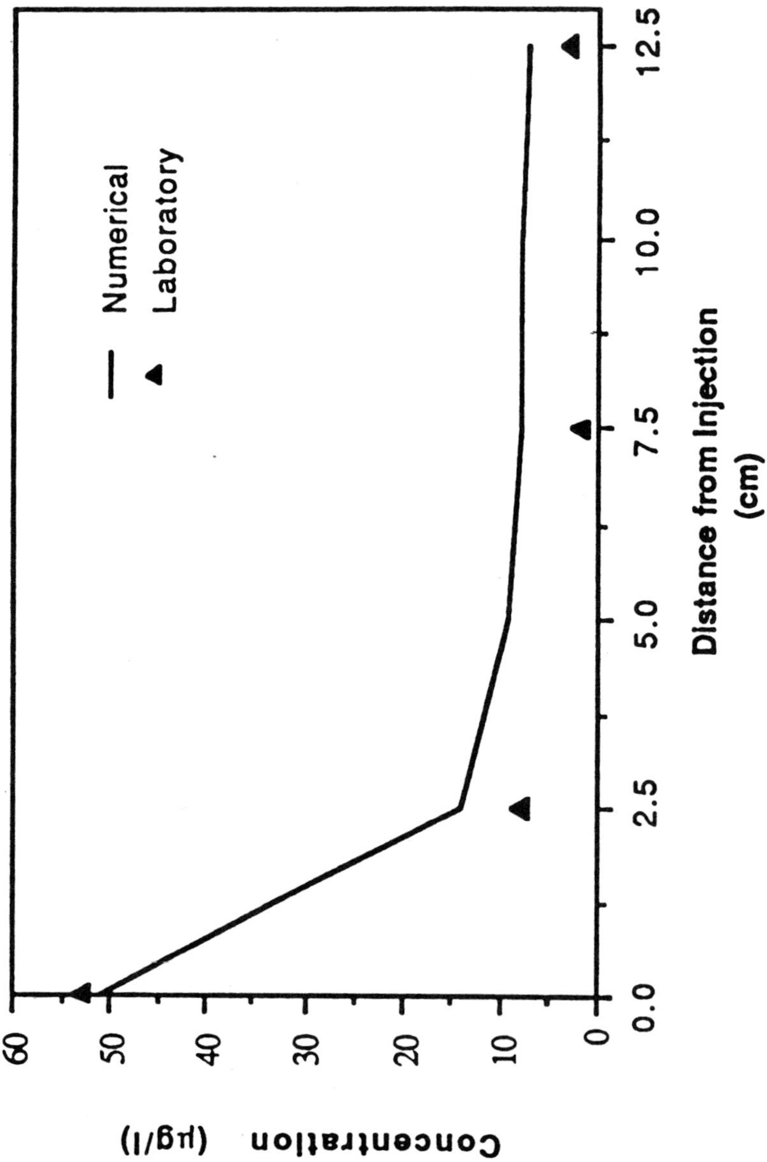

Figure 6. Prediction of the CTC profile at a detention time of 500 min. and with $k = 0.030$ μg/mg cell-day and $K_s = 4.5$ μg/l.

Bromoform, Ethylene Dibromide, Tetrachloroethene, and Trichloroethene

The second secondary substrate modeled was bromoform (BF), which entered the column at a concentration of 106 μg/l at a detention time of 50 min. As opposed to CTC, BF had no detectable removal at the 50 min. detention time. This was due to insufficient contact time with the biomass, and as a result, the detention time of the column was increased. The 125-min. detention time had an influent concentration of BF of 57 μg/l, and BF removal was observed. In this case, the kinetic parameters, k and K_s, were found for this detention time. Figure 7 shows the results of the numerical fitting between the experimental values and the model results using k and K_s values of 0.013 μg/mg-day and 9.5 μg/l, respectively.

The 500-min. detention time column had an influent concentration of 54 μg/l of BF. Figure 8 shows the results of the numerical model using the kinetic parameters from the 125-min. detention time. The numerical model predicted a slightly lower plateau concentration than was measured in the laboratory. This difference in removals perhaps can be attributed to biofilm loss between the two sampling periods. The assumption made was that the biomass distribution remained constant throughout the duration of the secondary-utilization experiments, even though some biomass loss probably took place.

Modeling predictions for the 50-min. detention time gave a prediction of only 1.5% removal of BF (results not shown). This negligible predicted removal was consistent with the undetectable removal for the experiments. The kinetic parameters of tetrachloroethene (TeCE), ethylene dibromide (EDB), and trichloroethene (TCE) were measured at the intermediate detention time of 125 min. Again, there were no detectable removals of these compounds at the lowest detention time of 50 min. The numerical model was fit to the laboratory data to obtain the k and K_s values of each compound. These values are shown in Table 3. Because the numerical fit to laboratory data was similar to that shown for BF in Figure 7, these curves are not presented. The same trend of somewhat greater removals of substrate predicted by the numerical model than measured in the laboratory applied to the

TABLE 3

K_s AND k VALUES OF TeCE, EDB, AND TCE OBTAINED FROM NUMERICAL CURVE FITTING

Compound	K_s (mg/l)	k (mg/mg-day)
TeCE	8.0	0.01760
EDB	9.0	0.00900
TCE	8.5	0.00935

greater detention time results for these three compounds. The differences are hypothesized to be the result of biomass loss.

Simulation of Bioreclamation Strategies

There are several possible strategies that can be used to achieve maximum performance of a bioreclamation site. Two very interesting strategies are presented in this chapter in the form of hypothetical examples. One strategy is to minimize the clogging from biomass growth. A clogging problem is exacerbated by injection of an excess amount of electron acceptor through one well or port. If the electron-donor concentration is relatively large compared to that of the electron acceptor, clogging is likely to develop in a region close to the point of injection. In order to reduce the potential for clogging, lower concentrations of the electron acceptor can be added at several locations along the flow path.

The clogging potential can be demonstrated by presenting an example problem. A model problem compares one injection of nitrate at 10.0 mg/l to a multiple injection of an equivalent amount of nitrate. For this example, SOC is assumed to be present in excess, so that nitrate is assumed to limit the growth throughout the length of the column. The concentration of SOC is assumed to be 20.0 mg SOC/l as it enters the column. Kinetic parameters for denitrification limited by NO_3^-, found in Table 2, were used for the model problem. The reactor parameters used in this example are shown in Table 4.

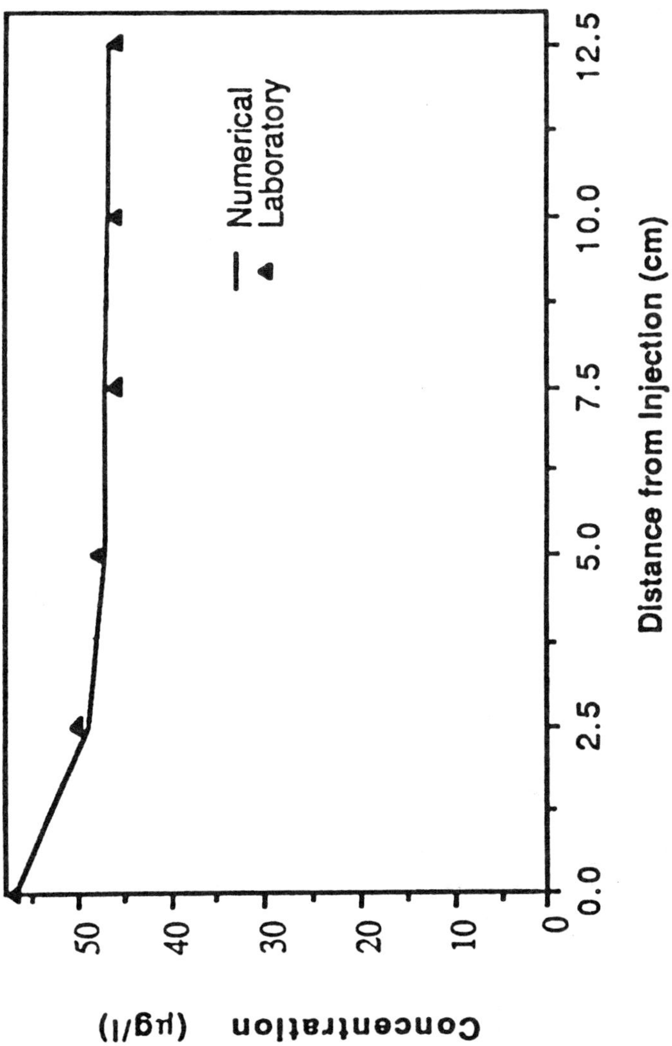

Figure 7. Numerical fit to the BF profile at a detention time of 125 min. The k and K_s are 0.013 μg/mg cell-day and 9.5 μg/l, respectively.

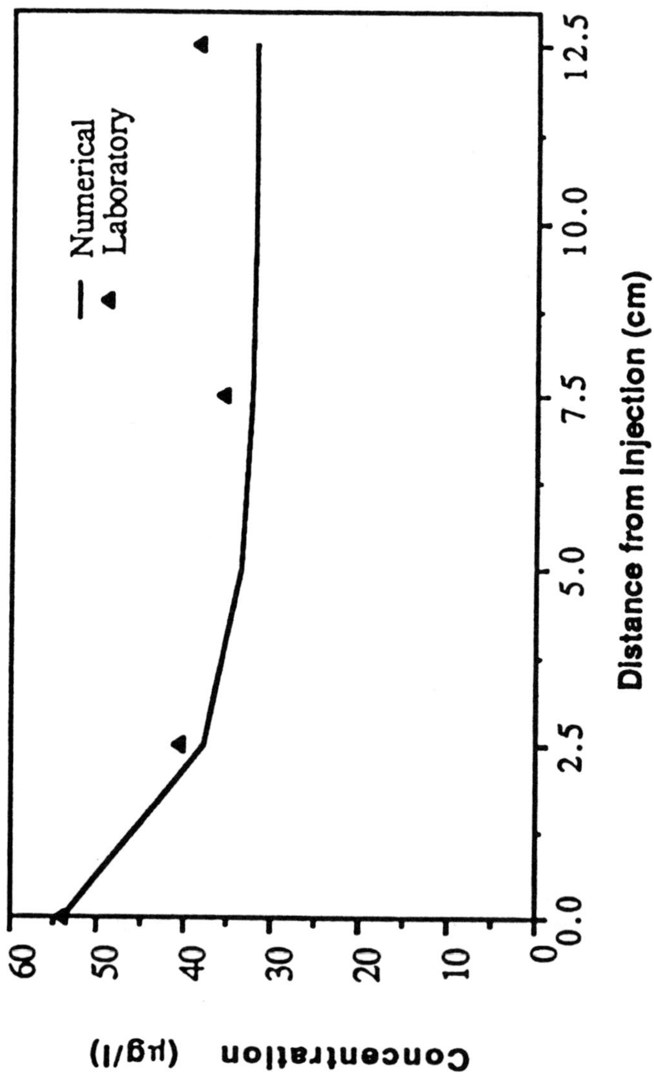

Figure 8. Prediction of the BF profile at a detention time of 500 min. and with k = 0.013 µg/mg cell-day and K_s = 9.5 µg/l.

TABLE 4

PARAMETERS USED IN CLOGGING EXAMPLE PROBLEM

Parameter	Units	Value
Nitrate	as NO_3^- -N	-
S^o	mg NO_3^-/l	3.3, 3.3, 3.3
L	cm	0.0289
v	cm/day	180.
ϵ	cm^3/cm^3	0.30
d_p	cm	0.12
a	cm^{-1}	50.0
Acetate	as SOC	-
S^o	mg SOC/l	20.00

Figure 9 shows the nitrate and SOC profiles when the NO_3^- is injected through one or three ports. It is evident that the net removal of SOC is nearly equivalent for either case, although it is slightly better with three injections. A relative measure of clogging potential can be estimated by considering the relative biofilm thickness throughout the length of the column. The relative biofilm thickness is the film thickness at a point in the reactor divided by the particle diameter. It is clear by comparing the relative biofilm thicknesses of the two scenarios, given in Figure 10, that the multiple injection gives much less potential for clogging.

The second strategy involves enhancing the removal of organic contaminants when the total available primary substrate is low. Injection of additional SOC when the original SOC is depleted should extend the BAZ and allow increased consumption of individual secondary substrates.

In order to demonstrate the advantage of an additional injection of carbon, a simple example was developed. Consider a situation such as the SOC profile in the one-BAZ column (see Figure 2). The SOC was rate limiting throughout the length of the column. The injected electron acceptor was removed approximately 50%. At this point, if additional acetate were added to the column, the BAZ could be extended. This configuration was modeled numerically using the laboratory kinetic

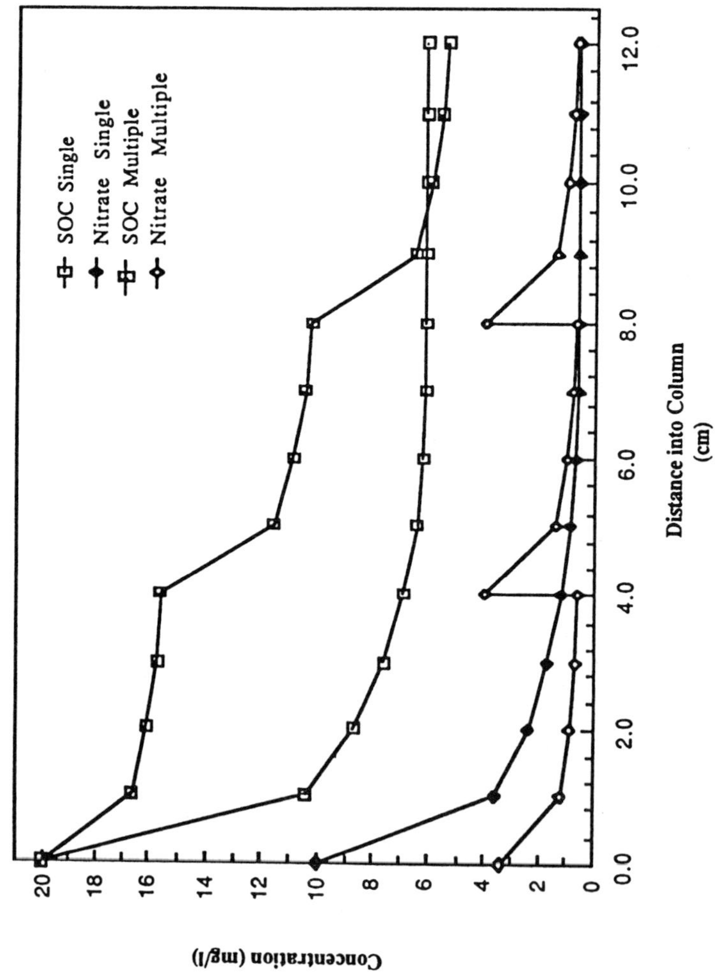

Figure 9. SOC and nitrate profiles for one and three injections of NO_3^-.

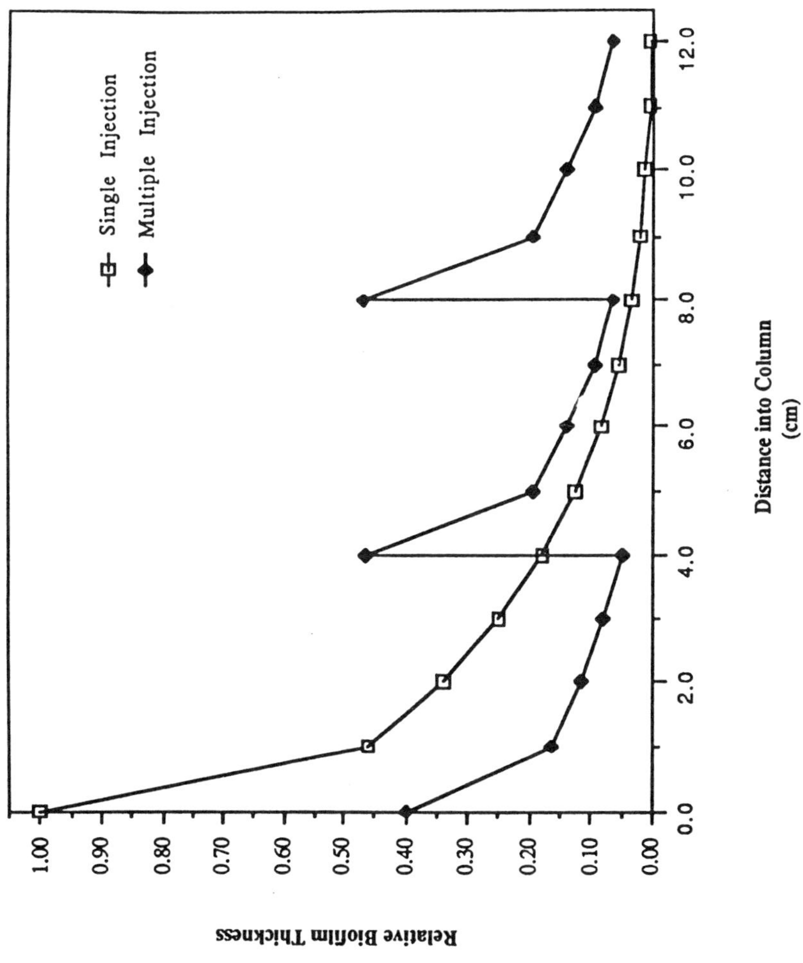

Figure 10. Relative biofilm thicknesses comparing single and multiple nitrate injections.

parameters. The initial concentration of nitrate was 10.0 mg NO_3^- -N/l, and the SOC was 6.5 mg SOC/l. At a distance of ten centimeters downstream from the nitrate injection, 10.0 mg SOC/l was injected. Figure 11 shows the resulting nitrate and SOC profiles. The BAZ was extended another 7.5 cm from the second injection.

The advantage of adding the SOC injection can be demonstrated further by considering the fate of CTC applied to the column at a concentration of 100 μg/l (using the same k and K_s values obtained in the section titled Carbon Tetrachloride). Figure 12 shows the profile of CTC throughout the length of the column. There is approximately 43% removal in the first ten centimeters of the column. An additional 40% removal is calculated for the last ten centimeters and is due to the injection of SOC. Thus, extension of the BAZ by addition of more biodegradable SOC significantly enhanced removal of CTC.

CONCLUSIONS

This research investigated the fundamental mechanisms that can act when an electron acceptor is injected along the flow path of an electron-donor-rich groundwater to establish a biologically active zone (BAZ) for degradation of pollutants that serve as primary and secondary substrates. The research methodology consisted of laboratory column experiments that were coupled with computer modeling.

The laboratory experiments demonstrated that lateral injection of NO_3^- could be successfully utilized to control the location and extent of BAZs in systems where acetate was fed as the sole carbon source. Columns containing one and two BAZs were successfully operated, and profiles of acetate and NO_3^- were determined. Additional measurements of steady-state biofilm thicknesses and densities gave further evidence of the value of lateral injection for spreading out biological activity along the flow path, which leads to enhanced biodegradation capability and diminished clogging potential. These experiments also demonstrated the deleterious effects of N_2 gas accumulation; N_2 gas bubbles that occurred as a result of denitrification tended to accumulate in the BAZs, resulting in reduced liquid contact times and lowered acetate removal efficiencies.

Laboratory experiments evaluating the secondary utilization of eight trace-concentration halogenated solvents were also conducted. Results of these experiments indicate that carbon tetrachloride was removed most

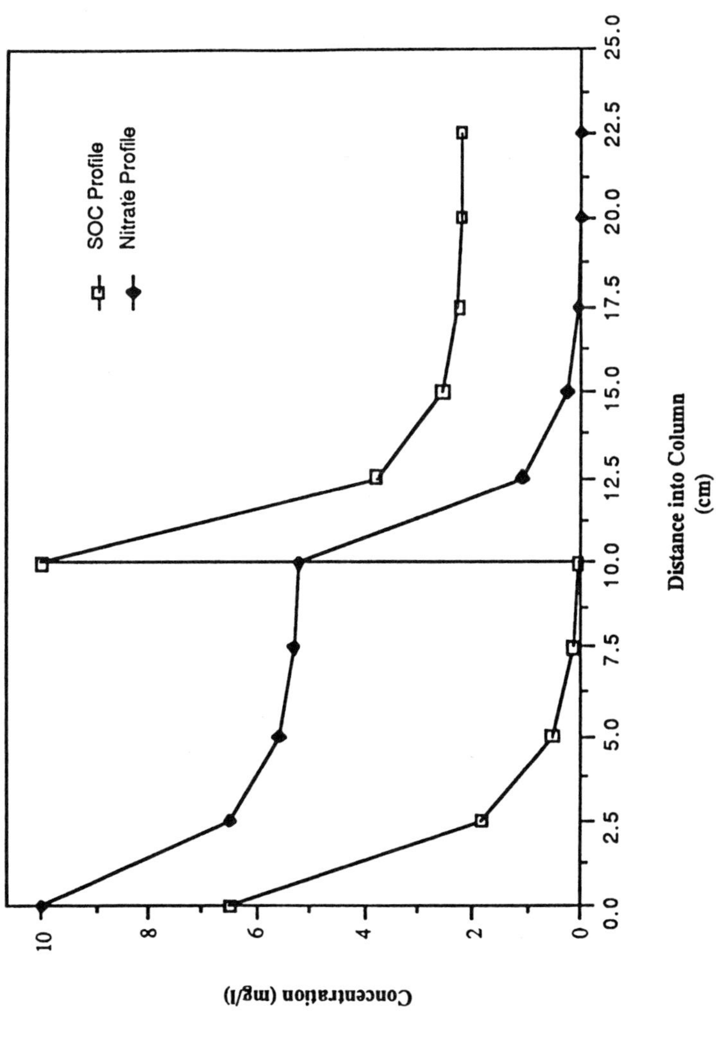

Figure 11. Profiles of SOC and nitrate after being injected alternately.

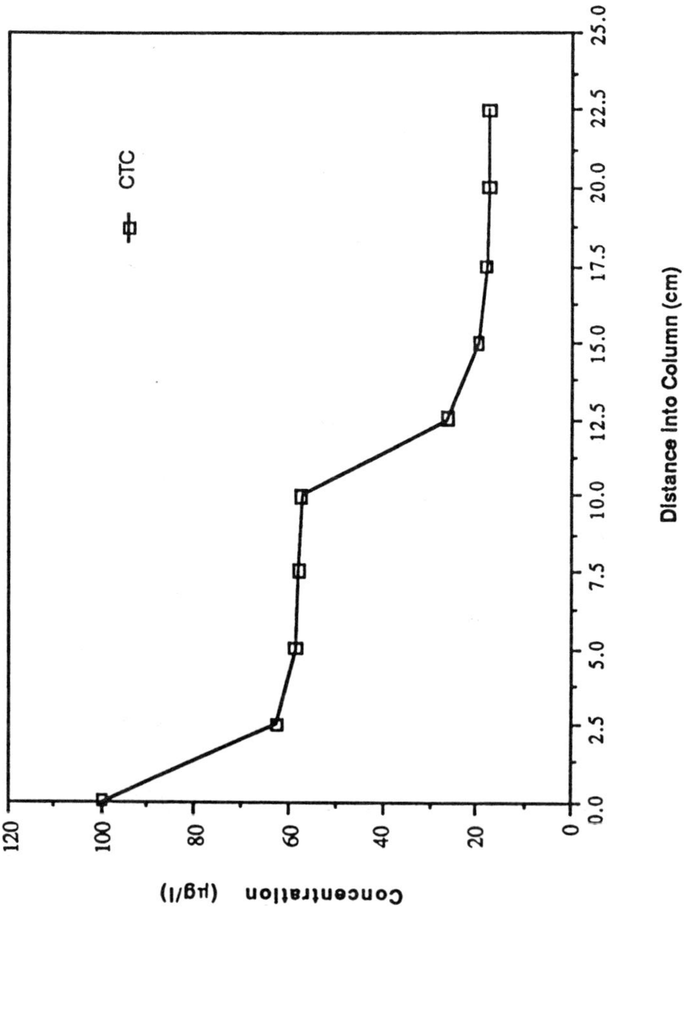

Figure 12. Profile showing additional secondary utilization of CTC after SOC is added by a second injection at 10 cm.

completely by denitrifying BAZs, while tetrachloroethene, bromoform, dibromoethane, and trichloroethene were removed to lesser degrees. Trichloroethane removal was slight. A significant result was that 1,2 and 1,3-dichlorobenzene were 20-30% removed; these compounds have previously been considered refractory under denitrifying conditions.

A highly efficient numerical model that couples solute transport mechanisms and biofilm kinetics was developed. Employing a quasilinearization technique for the biofilm reaction term, the model is capable of solving directly for the steady-state profiles of limiting substrate, biofilm thickness, non-limiting substrates, and secondary substrates. The predictive ability of the model was successfully verified by simulating the results of the laboratory experiments using independently determined kinetic parameters. Independently determined kinetic parameters did not exist for the secondary substrates; in this case, one set of results from the column experiments was used to obtain a best-fit set of kinetic parameters, which were then used to predict the results for experiments conducted with different liquid flow velocities. The model predictions correctly described all trends. Absolute deviations between predicted and experimental results were very small for cases involving acetate and nitrate; systematic deviations for some of the secondary substrates occurred and probably were due to a loss of biomass during the experiments conducted at the higher detention times.

The steady-state models were applied to investigate possible strategies to be used in field bioreclamations. The use of multiple injection wells was studied for its ability to decrease aquifer clogging potential by spreading out the distance over which the limiting substrate is added. Modeling results verified that the strategy of multiple injections could reduce high densities of biofilm accumulation near the injection well. Also investigated was the strategy of adding a supplemental carbon source to extend the length of a BAZ. The modeling illustrated that such an extension of the BAZ could be accomplished and could result in longer contact times for a secondary substrate in the BAZ, thereby increasing the removal of the secondary substrate.

The results of this research demonstrate that injection of limiting substrates along the groundwater flow path is a viable means of establishing spatially distributed BAZs for enhanced in-situ bioreclamation. Trace-levels of hazardous secondary substrates can be degraded as groundwater flows through the BAZs. The phenomena of formation of BAZs and substrate utilization within BAZs can be quantitatively interpreted and predicted at the laboratory scale by rigorous mathematical models that couple principles of solute transport and biofilm kinetics.

An ultimate goal is to develop the fundamental understanding of coupled biological and hydrological processes to a level sufficiently great that field-scale in-situ bioreclamation systems can be designed reliably. A critical need is to extend the research reported here to include transient and multi-dimensional aspects. In particular, the following areas of additional research are recommended:

1. Use the combined experimental and modeling approach to study transient biofilm kinetics and dual-substrate limitation.

2. Examine the use of alternative electron acceptors to establish specialized BAZs capable of degrading specific pollutants.

3. Conduct further study of the basic mechanisms of dichlorobenzene degradation under denitrifying conditions.

4. Examine the fundamental mechanisms of bacterial transport and attachment and their role in the establishment and extension of BAZs.

5. Study the effect of biological activity upon the hydraulic properties of aquifers.

6. Extend the computer modeling to consider transient, heterogeneous, multi-dimensional flow fields, as well as transient biofilms.

7. Study the phenomena controlling the biodegradation of organic contaminants which are strongly adsorbed or which form a nonaqueous phase.

Additional references for the reader to refer to are cited at the end of this chapter.

REFERENCES

1. Rittmann, B.E., A.J. Volocchi, J.E. Odencrantz, and W. Bae. "In-Situ Bioreclamation of Contaminated Groundwater," Illinois Dept. of Energy and Natural Resources, Report HWRIC RR-031, October, 1990.

2. American Public Health Association, Inc. "Standard Methods for the Examination of Waste and Wastewater." 15th ed. Washington D.C., 1981.

3. Atkinson, B., and S.Y. How. "The Overall Rate of Substrate Uptake (Reaction) by Microbial Films. Part II. Effect of Concentration and Thickness with Mixed Microbial Films." *Transactions, Inst. Chem. Engrs.*, 1974, (52) 260-272.

4. Bouwer, E.J. "Biotransformations of Contaminants in Groundwater." Paper presented at the Third National Water Conference, Philadelphia, PA, January, 1987.

5. Bouwer, E.J., and P.L. McCarty. "Transformation of Halogenated Organic Compounds Under Denitrification Conditions." *Appl. Environ. Microb.*, 1983, (45) 1295-1299.

6. Bouwer, E.J., and P.L. McCarty. "Utilization Rates of Trace Halogenated Organic Compounds in Acetate-Grown Biofilms." *Biotech. and Bioengr.*, 1985, (27) 1564-1571.

7. Casolari, A. "Microbial Death." In *Physiological Models in Microbiology, Vol. II*, edited by M.J. Bazin and J.I. Prosser, Boca Raton, FL: CRC Press, Inc., 1988, 1-44.

8. Chang, H.T. Mathematical Modeling and Scanning Electron Microscopic Study of Biofilms on Adsorptive Media. Ph.D. Dissertation. Department of Civil Engineering, University of Illinois, Urbana, IL, 1985.

9. Daugherty, R.L., and J.B. Franzini. *Fluid Mechanics with Engineering Applications*. New York: McGraw-Hill, 1977.

10. Flathman, P.E. et al. "Biological Spill Cleanup." *Proc. Technical Seminar on Chemical Spills*, October 25-27, 1983; Toronto, 117-130.

11. Flathman, P.E., J.R. Quince, and L.S. Bottomley. "Biological Treatment of Ethylene Glycol-Contaminated Groundwater at Naval Air Engineering Center, Lakehurst, New Jersey. *Proc. 4th National Symposium on Aquifer Restoration and Ground Water Monitoring*, May 23-25, 1984; Columbus, Ohio, 23-25.

12. Freeze, R.A., and J.A. Cherry. *Groundwater*. Englewood Cliffs, NJ: Prentice-Hall., 1979.

13. Ghiorse, W.C., and D.L. Balkwill. "Enumeration and Morphological Characterization of Bacteria Indigenous to Subsurface Environments." *Dev. Ind. Microb.*, 1983, 24.

14. Kobayashi, H., and B.E. Rittmann. "Microbial Removal of Hazardous Organic Compounds." *Environ. Sci. Technol.*, 1982, (16) 170A-181A.

15. Kuhn, E.P., P.J. Colberg, J.L. Schnoor, O. Wanner, A.F.B. Zehnder, and R.P. Schwarzenbach. "Microbial Transformations of Substituted Benzenes during Infiltration of River Water to Groundwater: Laboratory Column Studies." *Environ. Sci. Technol.*, 1985, (19) 961-968.

16. Lapidus, L., and G.F. Pinder. *Numerical Solution of Partial Differential Equation in Science and Engineering*. New York: J. Wiley and Sons, Inc, 1982.

17. Lee, E.S. *Quasilinearization and Invariant Imbedding*. New York: Academic Press, 1968.

18. Lyman, W.J., W.F. Reehl, and D.H. Rosenblatt. *Handbook of Chemical Property Estimation Methods*. New York: McGraw-Hill Book Co., 1982.

19. McCarty, P.L. "Energetics and Bacterial Growth," In *Organic Compounds in Aquatic Environments*, edited by S.D. Faust and J.V. Hunter. New York: Marcel Dekker, Inc., 1972.

20. McCarty, P.L., B.E. Rittmann, and M. Reinhard. "Trace Organics in Groundwater," *Environ. Sci. Technol.*, 1981, (15) 40-51.

21. Miller, M.M., S.P. Wasik, G.-L. Huang, W.-Y. Shiu, and D. Mackay. "Relationship between Octanol-Water Partition Coefficient and Aqueous Solubility," *Environ. Sci. Technol.*, 1985, (19) 522-529.

22. Nagel, G., W. Kuehn, P. Werner, and H. Sontheimer. "Sanitation of Groundwater by Infiltration of Ozone Treated Water," *GWF-Wasser/Abwasser*, 1982, (123) 399-407.

23. Namkung, E. Kinetics and Mechanisms of Low-Concentration Multisubstrate Utilization by Biofilms. Ph.D. Dissertation. Department of Civil Engineering, University of Illinois, Urbana, IL, 1985.

24. Namkung, E. and B.E. Rittmann. "Soluble Microbial Products (SMP) Formation Kinetics by Biofilms," *Water Research*, 1986, (20) 795-806.

25. Namkung, E., R.G. Stratton, and B.E. Rittmann. "Predicting Removal of Trace-Organic Compounds by Biofilms," *J. Water Poll. Control Fed.*, 1983, (55) 1366-1372.

26. Parker, G.A. In *Colorimetric Determination of Nonmetals*, edited by D.R. Boltz and Z.A. Howell. New York: J. Wiley and Sons, 1928.

27. Rittmann, B.E. "Comparative Performance of Biofilm Reactor Types," *Biotech. and Bioengr.*, 1982a, (24) 1341-1370.

28. Rittmann, B.E. "The Effect of Shear Loss on Biofilm Loss Rate," *Biotech. and Bioengr.*, 1982b, (24) 501-506.

29. Rittmann, B.E., and W.E. Langeland. "Simultaneous Denitrification with Nitrification in Single Channel Oxidation Ditches," *J. Water Poll. Control Fed.*, 1985, (57) 300-308.

30. Rittmann, B.E., and P.L. McCarty. "Evaluation of Steady-State Biofilm Kinetics," *Biotech. and Bioengr.*, 1980a, (22) 2359-2373.

31. Rittmann, B.E., and P.L. McCarty. "Model of Steady-State Biofilm Kinetics," *Biotech. and Bioengr.*, 1980b, (22) 2343-2357.

32. Rittmann, B.E., and P.L. McCarty. "Substrate Flux into Biofilms of any Thickness," *J. Environ. Engr.*, 1981, (107) 831-849.

33. Rittmann, B.E., P.L. McCarty, and P.V. Roberts. "Trace-Organics Biodegradation in Aquifer Recharge," *Ground Water*, 1980, (18) 236-243.

34. Saez, P.B., and B.E. Rittmann. "An Improved Pseudo-Analytical Solution for Steady-State Biofilm Kinetics," *Biotech. and Bioengr.*, 1988, (32) 362-368.

35. Stratton, R., E. Namkung, and B.E. Rittmann. "Biodegradation of Trace-Organic Compounds by Biofilms on Porous Media," *J. Amer. Water Works Assn.*, 1983, (75) 463-469.

36. Werner, P. "A New Way for the Decontamination of Polluted Aquifers by Biodegradation," *Water Supply*, 1985, (3) 41-47.

INDEX

O

P

U

V

W

X

Z